T0255858

Student Solutions Manual

Julie Clark
Hollins University

to accompany

Workshop Statistics
Discovery with Data
Fourth Edition

Allan J. Rossman
Beth L. Chance

California Polytechnic State University
San Luis Obispo

WILEY

John Wiley & Sons, Inc.

Founded in 1807, John Wiley & Sons, Inc. has been a valued source of knowledge and understanding for more than 200 years, helping people around the world meet their needs and fulfill their aspirations. Our company is built on a foundation of principles that include responsibility to the communities we serve and where we live and work. In 2008, we launched a Corporate Citizenship Initiative, a global effort to address the environmental, social, economic, and ethical challenges we face in our business. Among the issues we are addressing are carbon impact, paper specifications and procurement, ethical conduct within our business and among our vendors, and community and charitable support. For more information, please visit our website: www.wiley.com/go/citizenship.

Copyright © 2012   John Wiley & Sons, Inc. All rights reserved. No part of this publication may be reproduced, stored in a retrieval system, or transmitted in any form or by any means, electronic, mechanical, photocopying, recording, scanning or otherwise, except as permitted under Sections 107 or 108 of the 1976 United States Copyright Act, without either the prior written permission of the Publisher, or authorization through payment of the appropriate per-copy fee to the Copyright Clearance Center, Inc., 222 Rosewood Drive, Danvers, MA 01923 (Web site: www.copyright.com). Requests to the Publisher for permission should be addressed to the Permissions Department, John Wiley & Sons, Inc., 111 River Street, Hoboken, NJ 07030-5774, (201) 748-6011, fax (201) 748-6008, or online at: www.wiley.com/go/permissions.

ISBN 978-0-470-54726-7

Printed in the United States of America

10 9 8 7 6 5 4 3 2 1

Contents

Unit 1

Collecting Data and Drawing Conclusions

Topic 1

Data and Variables

Odd- Numbered Exercise Solutions

Exercise 1-7: Miscellany

a. Binary categorical; observational units: rolls of toilet paper

b. Binary categorical; observational units: applicants to graduate school

c. Quantitative; observational units: college students

d. Categorical; observational units: person

e. Binary categorical; observational units: person

f. Binary categorical; observational units: participant in sports

g. Quantitative; observational units: sports

h. Quantitative; observational units: the fifty states

i. Quantitative; observational units: baseball games or baseball stadiums

j. Quantitative; observational units: person

k. Quantitative; observational units: brides

l. Categorical; observational units: wedding couples

m. Quantitative; observational units: wedding couples

Exercise 1-9: Credit Card Usage

a. *Year in school*: categorical

   Whether or not the student has a credit card: binary categorical

   Outstanding balance on the credit card: quantitative

   Whether or not the outstanding balance exceeds $1000: binary categorical

   Source for selecting a credit card: categorical

   Region of the country: categorical

b. Answers will vary, but some sample questions include:

   Which class (freshman, sophomore, ...,) tends to have the largest outstanding credit card balance?

   Do all regions of the country tend to obtain their credit cards from the same source?

Exercise 1-11: Proximity to the Teacher

a. The observational units are the students.

b. One variable is the *quiz score*. This variable is quantitative. The other variable is the *distance the student is sitting from the teacher*. This variable is categorical if the *distance* is measured by the row in which a student sits. This variable is categorical and binary if the *distance* is thought of as "close to the teacher" and "far from the teacher". This variable would be quantitative if *distance* was the straight-line distance between the student and the teacher's typical position in the classroom.

Exercise 1-13: Candy Colors

a. The observational units are the pieces of candy.

b. The variable is the *color of the candy*. This variable is categorical (non-binary).

c. Now the observational units are the samples of 25 pieces of candy.

d. The variable is the *proportion of the sample that is colored orange*. This variable is quantitative.

Exercise 1-15: Children's Television Viewing

a. The observational units are the third and fourth grade students in San Jose.

b. The quantitative variables are *body mass index, triceps skinfold thickness, waist circumference, waist-to-hip ratio, weekly time spent watching television, weekly time spent*, and *weekly time spent playing video games*.

The categorical variables are *which school the student attends* and *gender*.

Exercise 1-17: Oscar Winners and Super Bowls

a. Answers will vary from student to student. Some examples include:

Categorical variables:

What is the movie's genre? Did the picture also win an Academy Award for best director?

Quantitative variables:

What was the total length (in minutes) of the movie? What was the production cost of the movie? How much did the movie gross during its first weekend of release?

b. Answers will vary from student to student, but some examples include:

Categorical variables:

In what city was the game played? Was the game played indoors or outdoors? Which conference was the winning team a member of? Was either team a wild card? Did the winner of the coin toss win the game?

Quantitative variables:

By how many points did the winning team win? How many people attended the game? What was the season percentage of wins for the winning team? What was the average total payroll for the winning team?

Exercise 1-19: Organ Donation

a. The observational units are the subjects in this study.

b. One variable is *whether the subject was told that the default option is to be a donor or not be a donor.* This variable is categorical and binary. The other variable is *whether the subject indicated that he/she was willing to be an organ donor or not.* This variable is categorical and binary.

c. *Does the default option on a state's driver's license application affect a citizen's willingness to become an organ donor?*

Exercise 1-21: Car Ages

a. The observational units would be the cars.

b. Variable 1 = *Whether the car is driven by a faculty member or a student* (binary categorical)

Variable 2 = *Age of the car* (quantitative)

c. This is not a variable; it is the research question under investigation rather than a measurement or category recorded about the individual cars.

Topic 2

Data and Distributions

Odd- Numbered Exercise Solutions

Exercise 2-7: Student Data

How many hours you slept: dotplot

Whether or not you have slept for at least 7 hours in the past 24: bar graph

Number of Harry Potter books read: dotplot

How many states you have visited: dotplot

Handedness: bar graph

Political viewpoint: bar graph

Day of the week on which you were born: bar graph

Average study time per week: dotplot

Number of birthday card received: dotplot

Gender: bar graph

Exercise 2-9: Value of Statistics

a. Answers will vary from student to student.

b. Answers will vary from class to class, but here are some sample answers.

Rating	1	2	3	4	5	6	7	8	9
Tally	0	0	1	0	5	6	11	6	6

c. Yes; seven was chosen more often than any other value.

d. Twenty-nine students (29/35, or 0.829) gave a response greater than 5; One student (1/35, or 0.029) gave a response less than 5.

e. The vast majority of this class, (more than 80%), feel that statistics is important to society. If fact, more than 65% of the class feel that statistics is *very* important to society. About 14% of the class is neutral about the important of statistics, and only 1 of the 35 students in this group believe that statistics is unimportant to society.

Exercise 2-11: Quiz Scores

a. Many answers are possible. Here are some examples followed by dotplots for all 4 quizzes shown on the same scale:

Quiz 1: 0, 1, 1, 2, 8, 8, 8, 9, 9, 9, 9, 9, 9, 9, 9, 10, 10, 10, 10, 10.

b. Quiz 2: 3, 4, 4, 4, 4, 5, 5, 5, 5, 5, 5, 5, 5, 5, 6, 6, 6, 6, 6, 6.

c. Quiz 3: 0, 1, 1, 2, 2, 3, 3, 4, 5, 5, 6, 6, 7, 7, 8, 8, 9, 10, 10, 10.

d. Quiz 4: 0, 0, 0, 1, 1, 1, 1, 1, 2, 2, 8, 8, 9, 9, 9, 9, 10, 10, 10, 10.

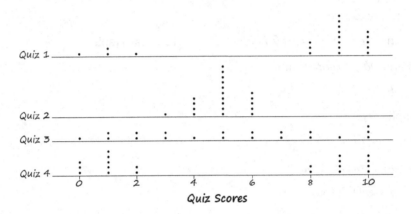

Exercise 2-13: Backpack Weights

a. Yes, it appears that males tend to carry slightly more weight in their backpacks than females. This is shown primarily by the centers in the dotplots. The graph for males appears shifted to the right of the graph for females.

b. No, it does not appear that one sex tends to carry a higher ratio of their body weight in their backpacks than the other sex. Both dotplots look quite similar in terms of shape, center and spread.

c. Males tend to weigh more than females, and so tend to carry more weight in their backpacks. But this factor is accounted for when you compute the ratio of backpack weight to bodyweight, as the ratio carried by each gender tends to be about the same.

Exercise 2-15: Highest Peaks

The highest points in the East tend to be significantly less high than those in the West. The elevations of the tallest peak in the East are all below 7000 ft, whereas more than half of those in the West are above 9000 ft. One state in the West is a high outlier with a peak of nearly 21,000 ft, and there is a cluster of about 11 western states with highest elevations between 12,000 and 15,000 ft.

Exercise 2-17: Roller Coasters

a. The observational units are the roller coasters.

b. The quantitative variables are *height, length, speed,* and *number of inversions*. The categorical variables are *type* (wooden or steel, binary), and *design* (sit down, stand up, inverted).

c. The typical height for a steel coaster is 148 ft.; the typical height for a wooden coaster is 100 ft.

d. The steel coasters *tend* to be taller than the wooden. Most of the steel coasters are taller than most of the wooden coasters and the typical steel coaster is taller than the typical wooden coaster.

e. No, the steel coasters are not always taller than the wooden coaster. There are some very short steel coasters and some relatively tall wooden coasters.

Exercise 2-19: Candy Colors

a. The observation units are the pieces of candy; the variable is the *color of the candy*. This variable is categorical.

b. Here is the bar graph:

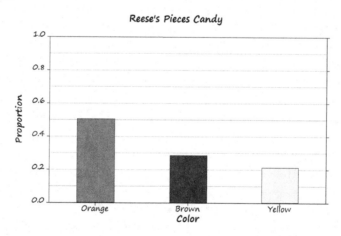

More than half of the candies were colored orange, just over a fourth (28%) were brown, while only 21% of them were yellow. This suggests that Hershey does not make equal proportions of each color.

c. Answers will vary by class.

Exercise 2-21: Coffee Consumption

a. Roughly 50% of the males almost never drink coffee, about 31% claim to sometimes drink coffee, and the remaining 19% drink coffee every day.

b. About 48% of the female students drink coffee every day, roughly 31% drink coffee sometimes and roughly 21% of the females almost never drink coffee.

c. Yes, these data indicate that males and females (at least in this introductory statistics class) differ with regard to coffee consumption. The majority of the males drink coffee never or occasionally, but almost half of the females reportedly drink coffee every day.

Exercise 2-23: Find your own

Answers will vary.

Topic 3

Drawing Conclusions from Studies

Odd- Numbered Exercise Solutions

Exercise 3-7: Student Data

a. Answers will vary by school and class.

b. Answers will vary by school and class.

c. Answers will vary by school and class.

Exercise 3-9: Community Ages

a. This number is a parameter as it would be a value describing the entire population of interest.

b. Yes, this sampling method would be biased. It would probably result in an overestimate the average age of residents as younger residents do not attend church as frequently as older residents do.

c. Yes, this would be a biased sampling method. This method would underestimate the average age of residents because the drivers at the daycare facility tend to be young adults, not middle-aged or elderly. This method would also exclude all residents who are not yet old enough to drive.

Exercise 3-11: Class Engagement

a. No; this is an observational study and there are at least two potential confounding variables that could explain the higher level of engagement in the statistics class. So you cannot attribute the difference in engagement to the subject matter.

b. One confounding variable is the *time of the class*. Perhaps more students are in attendance and are more alert at 11:00 AM than at 8:00 AM, and therefore more engaged in class. The other confounding variable is *the instructor*. Perhaps one instructor has a more interactive teaching style than the other, and this might be the cause of the increased interaction from the students.

Exercise 3-13: Alternative Medicine

This sample result is probably not representative of the truth concerning the population of all adult Americans because the sampling method is biased. Only readers of *Self* magazine were part of the poll, and the readers of this health magazine were probably more likely to be the type of people who are willing to try alternative medicines than the non-readers (which is a result of a bad sampling frame). Therefore, this result is very likely to be an overestimate of the proportion of all adult Americans who have used alternative medicines.

Exercise 3-15: Junior Golfer Survey

a. No, this is not likely to be a representative sample of all American teenagers because most teenagers do not play golf.

b. Yes, this sampling procedure is likely to be biased with respect to voting preference. Golfing is an expensive sport, and the wealthy tend to vote Republican, so the teenagers who were sampled may have tended to grow-up in Republican households.

c. The following graph displays the responses:

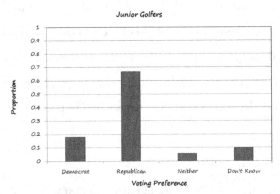

This graph shows that the majority of respondents indicated they were more likely to vote for a Republican. If you don't believe most teenagers are Republicans, this gives you evidence that the sampling method is over-representing the Republicans in the population.

d. Yes, this sampling procedure is likely to be biased with regard to both of these variables. If junior golfers tend to come from more affluent families, they almost certainly have a cell phone and computer in their home, making online access readily available and probably giving them more free time to spend on the computer. Of course, if they are more physically active and training for tournaments, they might tend to spend less time online than a typical teen.

Exercise 3-17: Foreign Language Study

a. Yes, these are observational studies. Researchers could only have passively observed the association between foreign language study and verbal SAT scores rather than determining for students whether they took foreign language in high school.

b. No, it is not legitimate to conclude that foreign language study causes an improvement in students' verbal abilities because this is an observational study with potential confounding variables. One possible confounding variable is verbal aptitude. Perhaps students with strong verbal aptitudes choose to enroll in foreign language courses and also perform well on the verbal portion of the SAT exam. Students with weaker verbal skills may avoid foreign language courses and may also perform less well on the verbal portion of the SAT.

Exercise 3-19: Smoking and Lung Cancer

a. The explanatory variable is *what are the subject's smoking habits?*. The response variable is *whether or not the subject died of lung cancer*.

b. Yes, this is an observational study. The researchers passively observed the smoking habits and life spans of their subjects rather than actively imposing the smoking habits on the individuals.

c. Yes, you should have qualms about generalizing these results to a larger population. The subjects were all males and were haphazardly selected by volunteers, so the results definitely should not be extended to women. The results might also be unrepresentative of the general male population as well, depending on how the volunteers selected the individuals.

Exercise 3-21: Buckle Up!

a. Yes, this is an observational study because you used existing data about the states. Researchers could not impose the treatment of "wearing a seatbelt" on some subjects.

b. No, you cannot conclude that the tougher seatbelt laws cause a higher proportion of residents to comply because this is an observational study.

Exercise 3-23: Pet Therapy

a. Yes, this is an observational study because you are passively observing and recording information about the patients instead of randomly determining which individuals own a pet.

b. The explanatory variable is *whether or not a recovering heart attack patient has a pet*. This variable is categorical and binary. The response variable *is whether or not the patient survives for five years*. This variable is categorical and binary.

c. No, you cannot conclude that pet ownership leads to therapeutic benefits for heart attack patients based on this study, because it is an observational study and you can never conclude cause-and-effect from an observational study. There are many potential confounding variables such as the *amount of daily exercise that the heart attack victim gets* that could explain the association we observed.

Exercise 3-25: Pursuit of Happiness

No, these study results do not establish a causal connection between income and happiness because this is an observational study and you cannot conclude cause-and-effect from an observational study. There are many potential confounding variables, such as *type of employment*, which could explain the association.

Exercise 3-27: Parking Meter Reliability

If the meters were *randomly* selected from Berkeley, it would be reasonable to generalize these results to all Berkeley parking meters. However, because they were not selected from all California parking meters, you wouldn't be willing to generalize the results to this population.

Exercise 3-29: Pulling All-Nighters

a. The explanatory variable is *whether or not the student claimed to have pulled an all-nighter*, and it is a binary categorical variable. The response variable is the *GPA*, and it is a quantitative variable.

b. Both 3.1 and 2.9 are statistics because they were calculated from the sample of 120 students at St. Lawrence University.

c. No, you cannot legitimately conclude that pulling all-nighters causes a student's GPA to decrease because this is an observational study and so there are potential confounding variables that could account for the apparent association. For example, perhaps students who claim to pull all-nighters do so because they do not keep-up in their courses, and therefore are not well prepared for the exams. This lack of preparation might be the cause of the lower GPAs.

d. You could reasonably generalize these results to all students at St. Lawrence University and perhaps to other small liberal arts colleges in the Northeast if the 120 students in this sample were randomly selected from all students at the University. But it would not be wise to generalize beyond these populations.

Exercise 3-31: Candy Cigarettes

a. The explanatory variable is *whether or not a subject used candy cigarettes as a child*, and this is a binary categorical variable. The response variable is *whether or not the subject is a current smoker*, and it is also binary categorical.

b. A confounding variable is an undefined/unrecorded variable whose effects on the response variable are indistinguishable from the explanatory variable. The *parent's smoking status* would be a confounding variable if the children whose parents smoked tended to also be the same children who used candy cigarettes. It seems plausible that those parents would be more willing to have candy cigarettes around the house and that the children would want to emulate their parents. It may well be that seeing one's parents smoke is what encourages children to use candy cigarettes in the first place and then to choose to smoke later in life.

Exercise 3-33: Tattoos

a. The response variable is *whether or not an American adult has a tattoo*. This variable is categorical and binary.

b. The number 14% is a statistic because it was obtained from the sample of 2302 American adults.

c. Because this sample was self-selected (chosen from people who agreed to participate in Harris surveys), it is not likely to be representative of the population of American adults. At most you should generalize these findings to the population of adult American who have access to the Internet.

d. The parameter of interest is the percentage of all adult Americans who have a tattoo.

Exercise 3-35: Spending on Mother

a. The population of interest is all adult Americans.

b. The sample is the consumers who responded to the NRF survey. The sample size was 7859.

c. The values 84.5% and $139.14 are both statistics as they are obtained from the sample of consumers who responded to the survey.

d. One parameter is the *percentage of all adult Americans who planned to celebrate Mother's Day in 2007*. The other parameter is the *average amount that all adult Americans were expecting to spend on Mother's Day that year*.

e. It is important to know whether the 7859 consumers were randomly selected because this information enables you to determine whether or not you can generalize the results of this survey to all adult Americans. This information would tell you whether the given statistics are likely to be representative of the entire population, or if the sampling method is likely to be biased.

Exercise 3-37: Home Court Disadvantage?

The proportion of sell-outs against a playoff team from the previous season was 11/21, or 0.5238, whereas the proportion of sell-outs against a team that was not in the playoffs was 7/20, or 0.35. The Thunder's proportion of wins against playoff teams was 7/21, or 0.3333, whereas their proportion of wins against non-playoff teams was 8/20, or 0.4. Because the Thunder tended to play 'good' teams in their sell-out games, and were not as successful against these 'good' teams, *whether or not a team was a playoff team in the previous season* may well account for the apparent home-court disadvantage at the Thunder's games.

Topic 4

Random Sampling

Odd- Numbered Exercise Solutions

Exercise 4-7: Sampling Words

a. This variable is categorical and binary.

b. The proportion of the words that contain at least five letters is 99/268 or 0.369.

c. The answer to part b is a parameter because it is a value calculated using all 268 words in the population (Gettysburg Address).

d. No; because of sampling variability you would not expect the sample proportion to equal .369, but you would expect it to be reasonably close most of the time. (In fact, with a sample of size 5, the sample proportion could not equal 0.369; it could only be 0, .2, .4, .6, .8, or 1.)

Exercise 4-9: Sampling Senators

a. The observational units are the U.S. senators. The variable is the *years of service in the senate*. The population is the current 100 U.S. senators. The sample is the 5 selected current U.S. senators. The parameter is average years of service of all 100 U.S. senators. The statistic is average years of service of the 5 selected senators.

b. This sampling method would most likely overestimate the average years of service since your classmates would most likely select names of well-known senators who have been serving in the senate for a long time. (You also need to worry about a tendency for students to mention the senators from their own state more than those from other states.)

c. No, increasing the sample size will not correct for a biased sampling method. Students would still tend to over-represent the senators who have served longer.

d. Obtain a list of the current senators. Number each senator in the list from 00-99. Select any row of the Random Digits Table and read the row as a sequence of two-digit numbers. These two-digit numbers tell you which senators from your list will make up your sample. Continue selecting senators until you have five senators in your sample. Skip any repeated two-digit numbers.

e. Obtain a list of the current representatives. Number each representative in the list from 000-434. Select any row of the table of Random Digits Table and read the row as a sequence of three-digit numbers. These three-digit numbers tell you which senators from your list will make up your sample. Continue selecting representatives until you have ten representatives in your sample. Skip any repeated three-digit numbers, or numbers greater than 434. If necessary, continue to another row of the Random Digits Table.

Exercise 4-11: Rose-y Opinions

a. The observational units are the 1000 individuals leaving a Los Angeles Lakers' basketball game. The variable is *whether they have a favorable or unfavorable opinion of Pete Rose.* This is a binary categorical variable.

b. The population of interest is American sports fans. The sample is the first 1000 people leaving an LA Lakers' basketball game.

c. This was not a randomly selected sample. People attending this basketball game are not necessarily sports fans in general or may be extreme LA Lakers fans, or simply basketball fans. This is an example of convenience sampling and is unlikely to result in a representative sample.

d. No, the individuals in the sample may still be only interested in basketball and not sports in general.

e. If you have a list of subscribers to *Sports Illustrated* you could number the list and use a table of random digits or computer to select a random sample of subscribers. The population who would be represented by this sample would be all readers of *Sports Illustrated*, which would certainly be more representative of the general sports fan than the previous samples.

f. The parameter is the percentage is American sports fans who have an unfavorable opinion of Pete Rose. Its value is unknown. The statistic is the 49% of the 1000 people interviewed by the Gallup pollsters who said they had an unfavorable opinion of Pete Rose.

g. The value of the statistic would most likely change if Gallup had selected another random sample of 1000 people to interview, but the value of the parameter would remain the same.

Exercise 4-13: Sport Utility Vehicles

a. The observational units are the vehicles. The variable is *whether or not the vehicle is an SUV.* The population is all vehicles on the road in your hometown. The sample is the vehicles that pass by the intersection between 7 and 8 AM that morning. The parameter is the proportion of all vehicles in your hometown that are SUVs. The statistic is the proportion of all vehicles that pass by that morning that are SUVs.

b. The vehicles that you observed between 7 and 8 AM may not be representative of all vehicles on the road. For example, the vehicles may be used to carpool children to school, and therefore, over-represent larger families with multiple children and larger cars; or they may be predominately commuter vehicles rather than weekend recreational vehicles and therefore under-represent the proportion of SUVs.

c. The sampling frame is the list of cars sold by that dealer.

d. The recently purchased vehicles will probably not represent the vehicles on the road in your town. For example, there may have been a backlash against SUV's recently because of high gas prices so that fewer SUVs were purchased in the last year, yet many people would still own them from purchases made several years ago.

Exercise 4-15: Emotional Support

a. Hite's sampling method is likely to be biased in the direction of women who think they give more support than they receive. Her sample was self-selected with a response rate of less than 50%, and she sampled from women's groups where women are likely to join because they feel they aren't getting the kind of companionship they want from their husbands or boyfriends.

b. Hite's poll surveyed the larger number of women.

c. The ABC News/*Washington Post* poll was probably more representative of the truth about the population of all American women because they used random sampling that was presumably unbiased.

Exercise 4-17: Phone Book Gender

a. The parameter is the proportion of women living in San Luis Obispo County. The statistic is the proportion of women listed on the randomly selected phone book page.

b. This sampling technique will give a biased estimate for the proportion of women living in San Luis Obispo County because the phone listings of many married women are often only under their husbands' names. In addition, many single women choose not to list their phone numbers to avoid harassing phone calls. Therefore, you should expect the statistic will be an underestimate of the population parameter.

Exercise 4-19: Voter Turnout

a. The proportion of people who said they had voted is 1783/2613, or 0.682.

b. This is a statistic because it is a number calculated from a sample (of 2613 adults).

c. The following bar graph displays the proportions who claim to have voted and not voted:

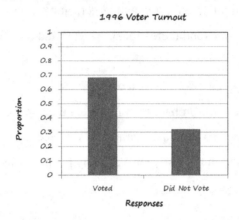

d. This number (49%) is a parameter because the Federal Election Commission has the records of all registered voters. Everyone who was eligible to vote was included in this number.

e. No, the sample grossly overestimated the proportion of eligible voters who actually voted.

f. Although the sample result is unlikely to match the population value exactly, this difference is probably too large to be attributed to sampling variability.

g. People may be reluctant to tell the truth (and seem unpatriotic) and so may overstate whether or not they voted. They might not remember that they didn't vote in this particular election. Even with random samples, you have to worry about the honesty of the respondents in surveys.

Activity 4-21: Prison Terms and Car Trips

a. Prisoners with longer terms have a higher probability of ending up in the sample (similar to how longer words are more likely to be selected when you point your finger at one spot on the page).

b. Cars engaged in longer trips have a higher chance of being observed at a particular time point than cars on shorter trips.

c. Many answers are possible, but one example is estimating the average length of time that people have been employed by a particular company. If you take a random sample of current employees, employees who have been around longer have a better chance of ending up in the sample.

Exercise 4-23: Waiting for Dentist

a. The population is all visits of all patients to your dentist's office during this year.

b. The number you are trying to estimate is a parameter because it is a number that summarizes the waiting time of all visits to your dentist's office during the year.

c. The average that you actually calculate will be a statistic because it is calculated from a sample of only some visits to your dentist's office.

d. This sampling method is still biased because patients with longer waiting times are more likely to be in the waiting room when you visit. There may be quite a few patients who are in and out quickly, but it will be harder to include them using this method. This sampling method will over-represent the longer waiting times.

Exercise 4-25: Pop vs. Soda

a. The most popular answer was "soda". The proportion of respondents who answered "soda" was 120130/293772 or 0.4089. This number is a statistic because was obtained from a sample.

b. This sampling technique is not likely to produce a sample that is representative of all Americans because the sample was self-selected, and restricted to people who visited the popvssoda.com website, and who felt strongly enough about the issue to vote.

c. No, the sample size of 293,772 does not correct for the biased sampling method.

Exercise 4-27: In the News

Answers will vary by student.

Topic 5

Designing Experiments

Odd- Numbered Exercise Solutions

Exercise 5-7: An Apple a Day

a. Anecdote

b. Observational study

c. Experiment

Exercise 5-9: Ice Cream Servings

a. The explanatory variables are the *large or small bowl* (binary categorical) and the *large or small scoop* (binary categorical). The response variable is *the amount of ice cream eaten* (quantitative).

b. This is an experiment because the researchers actively imposed the treatments on the subjects by randomly assigning the size of the scoops and bowls.

c. The random assignment was important because it controlled for the potential confounding variable of self-selection. If the nutrition experts were allowed to choose for themselves, those who tended to have small appetites might have chosen the smaller bowls and/or scoops and consequently eaten less ice cream. Then appetite would be confounded with bowl/scoop size.

d. The nutrition experts did not know that there were two different sizes of bowls and scoops being distributed, so they would not adjust the amount of ice-cream they ate in order to be more in line with one of the other groups.

e. Because this study was a well-designed, randomized, controlled experiment, it is valid to draw a cause-and-effect conclusion between size of bowl or scoop and size of the ice cream serving.

f. You have controlled for this potentially confounding variable by randomly assigning the subjects to the treatment groups. The only difference between the groups should be the bowl and scoop sizes.

Exercise 5-11: Foreign Language Study

a. No, you cannot conclude that foreign language study improves your verbal skills. Because this was an observational study, there are many confounding variables that could explain the association.

b. A controlled study would need to randomly assign students to different treatment groups (i.e. foreign language study and no foreign language study) and then later compare the verbal SAT scores of the two groups. This would ensure that hidden confounding variables such as verbal aptitude would balance out between the groups

c. It might not be feasible to carry out such an experiment because you generally cannot control which courses students do or do not take.

Exercise 5-13: Pet Therapy

a. The explanatory variable is *whether or not the heart attack patient owns a pet* (binary categorical). The response variable is *whether or not the patient survived for five years* (binary categorical).

b. This is an observational study. The researcher has passively observed and recorded information on pet ownership and the patient's recovery rather than assigning some people to own pets and others to not own pets.

c. Yes, there is a group of patients who do not own pets for comparison.

d. No, this study does not make use of randomization. Patients were not randomly selected or randomly assigned to treatment groups.

e. No, you cannot conclude that owning a pet has a therapeutic effect for heart attack survivors because there may be lurking variables that explain the association. You cannot conclude causation with an observational study.

f. This study could be a controlled experiment if the researcher used randomization to determine whether or not the patient owned a pet. In this case the researcher would *actively* impose the treatment on the subjects. The experimenter would then hope to see the direct effect of pet ownership on the recovery rate of heart attack patients.

g. This is debatable. Is it feasible to tell someone to own a pet? Probably not.

Exercise 5-15: Reducing Cold Durations

a. The experimental units are the 104 subjects reporting to the lab within 24 hours of getting a cold.

b. The explanatory variable is the *amount of zinc nasal spray* (full, low or no dosage). The response variable is the *duration of cold symptoms*.

c. This is an experiment because the researchers randomly assigned the subjects to the treatment groups (amount of zinc spray) and actively imposed the treatments on the patients.

d. The researchers used a placebo to ensure that if the subjects' colds improved because of *any* treatment, this effect would be seen equally in each of the groups.

Exercise 5-17: Natural Light and Achievement

a. Researchers would randomly assign the students to two different treatment groups – one with high natural light and one with low natural light. Then the researchers would compare the standardized test scores of the students in these two groups.

b. It would be difficult to carry out this experiment because there are ethical considerations that could prevent you from depriving students of natural light and also from possibly detrimentally affecting their education.

c. John B. Lyons could say "There is a causal relationship between daylight and achievement" if this was a well-designed, randomized, comparative experiment.

Exercise 5-19: Capital Punishment

a. No, this is not an experiment because the researcher did not impose the death penalty statute on the states that have it or prevent other states from having it.

b. No, you cannot conclude that the death penalty caused the difference in homicide rates because this is just an observational study. There may be confounding variables (such as the state's overall crime rate or legal system) that could also affect the response variable.

c. No, you cannot conclude a lack of causation either because this is just an observational study. There could be other variables that are masking the effect of the death penalty.

Exercise 5-21: Therapeutic Touch

a. This was an experiment. Emily imposed the treatment (her hand) on the subjects.

b. Emily flipped a coin to decide which of the subject's hands she would hold hers over.

c. This study was not double-blind. Emily was aware of which subjects received which treatments.

d. No; Emily's sample consisted of volunteers. It was not randomly selected from all practitioners.

e. No, you should not attribute this tendency to detection of Emily's energy field. Emily used only practitioners of therapeutic touch in her study, she did not have a control group of people who did not claim to participate in this practice with which to compare.

Exercise 5-23: Proximity to the Teacher

a. The observational units are the students. The explanatory variable is *whether the student sits close to /far away from the teacher*. The response variable is *performance on quizzes*.

b. Researchers would randomly assign the students to two different treatment groups – one that sits close to the teacher and that sits far from the teacher. Then the researchers would compare the quiz scores of the students in these two groups.

c. You will be able to conclude that sitting closer to the teacher does (or does not) cause students to perform better on quizzes if you use a well-designed, randomized, comparative experiment.

d. You would not have to worry about the ethics of assigning seats to students or possibly detrimentally affecting the students' education through their seat assignments.

Exercise 5-25: Dolphin Therapy

a. This is an experiment because the researcher actively imposed the assignment to the two groups (swimming with or without dolphins) on the subjects.

b. The explanatory variable is *swimming with dolphins or swimming/snorkeling without dolphins*. This variable is categorical and binary. The response variable is the *change in depression symptoms*. This variable is presumably categorical.

c. Assuming that the patients were randomly assigned to the two groups, yes, you can conclude that swimming with dolphins improves depression symptoms because this was a well-designed, randomized, comparative experiment.

d. No, the subjects were not blind as to which treatment they received. It would be impossible to achieve blindness in this experiment because you cannot make people unaware that they are swimming with dolphins.

Exercise 5-27: Friendly Observers

a. This study is an experiment because the researcher randomly assigned the subjects to the two groups (only participants would win $3, and participants plus observers would win $3).

b. The observational units are the subjects playing a video game.

c. The explanatory variable is *whether or not the observer was to share in the prize*. The response variable was *whether or not the threshold was beaten*.

d. This study makes use of blindness because the subjects were not told that there were two different groups, or into which group they were placed.

Exercise 5-29: Facebook and Grades

a. The explanatory variable is *whether or not students have a Facebook account*. It is categorical and binary.

b. The response variables are *the number of hours spend studying per week* (quantitative), and *grade point average* (quantitative).

c. This is an observational study because the researchers did not actively, randomly impose the explanatory variable (*having a Facebook account or not*) on some students.

d. It is not reasonable to conclude that having a Facebook account causes less studying or lower grades because this was an observational study and there are many potentially confounding variables that could account for the observed association.

Exercise 5-31: Bonus or Rebate

a. The explanatory variable is *whether the $50 was presented as a "tuition rebate" or as "bonus income."* This variable is categorical and binary.

b. One response variable is whether or not the undergraduate kept the entire $50 (binary categorical), and the other variable is *the amount of the $50 the undergraduate spent* (quantitative).

c. This is an experiment because the undergraduates were randomly assigned to one of the two treatment groups (tuition rebate or bonus income).

d. Because this was a well-designed, randomized, comparative experiment, and there was a significant difference in the amount saved by the two groups, you can legitimately conclude that the way the $50 is presented (tuition rebate or bonus income) is the cause of how much money is saved. This should be the only difference between the two groups.

e. You cannot safely generalize these results to all college students. All of the students in this sample were Harvard undergraduates. They are unlikely to be typical of U.S. college students.

Exercise 5-33: CPR on Pets

If you wished to conduct a study on this question in your community, you should use random sampling because it would be very difficult to use random assignment in this situation. You could not easily randomly assign people to own a cat or dog (or neither).

Exercise 5-35: In the News

Answers will vary.

Unit 2

Summarizing Data

Topic 6

Two-Way Tables

Odd- Numbered Exercise Solutions

Exercise 6-7: "Hella" Project

  a. This is an observational study.

b. The explanatory variable is *whether the student is from northern or southern California*. The response variable is *whether the student used "hella" in her or his everyday vocabulary*.

c. A two-way table of the responses is shown here:

	Northern Californians	Southern Californians	Total
Uses "Hella" Regularly	10	3	13
Does Not Use "Hella" Regularly	5	22	27
Total	15	25	40

d. For Southern Californians, the proportion of students who "hella" regularly is 3/25 or .12. For Northern Californians, the proportion is 10/15 or .667.

e. The following segmented bar graph displays these data:

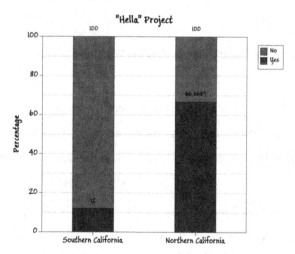

f. The data do seem to support the students' conjecture that Northern Californians are more likely to use the word "hella". The students in this sample from Northern California were more than five times as likely (relative risk = .6617/.12 = 5.56) to use "hella" in their everyday vocabulary as the students from Southern California.

Exercise 6-9: Suitability for Politics

The following segmented bar graph displays these results:

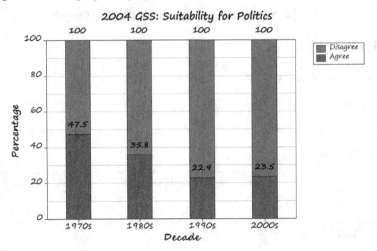

According to the data, in the 1970s, about 47% (2398/5049) of those polled agreed with this statement on suitability for politics, and this percentage declined to about 36% (2563/7160) in the 1980s, about 23% in the 1990s (1909/8336) and about 23.5% in the 2000s (802/3411). Therefore, people have tended to disagree more and more with this statement until the turn of the century, when opinions may have leveled-off slightly.

Exercise 6-11: Children's Television Advertisements

a. This is a 5 × 3 table.

b. The proportion of food advertisements on BET that were for fast food is 61/162 ≈ 0.377.

c. The proportion of fast-food advertisements that were on BET is 61/93 ≈ 0.656.

d. Here is the conditional distribution of the types of food commercials shown:

	BET	WB	Disney
Fast Food	0.377	0.386	0.000
Drinks	0.407	.0108	0.455
Snacks	0.019	0.000	0.182
Cereal	0.093	0.193	0.364

Candy	0.105	0.313	0.000

e. The following segmented bar graph displays these conditional distributions:

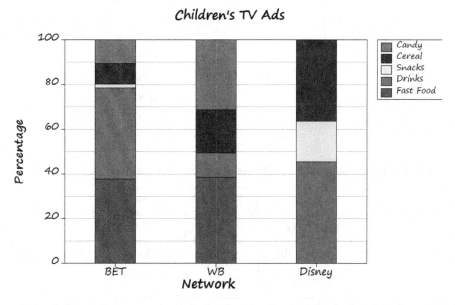

Children's TV Ads

f. The Disney channel showed no commercials for fast food or candy during this time period (in fact, they showed very few food advertisements at all). About 37% of the BET and WB food advertisements were for fast food and almost none of them were for snacks. The percentage of food advertisements on the WB network for cereal was more than double that of the BET network (19% vs. 9%) and for candy the percentage was tripled (31% vs. 10%).

Exercise 6-13: Weighty Feelings

a. The explanatory variables is *gender*. The response variable is *feelings about one's weight*.

b. Here is the marginal distribution for the variable *feelings about one's weight* (the variable *gender* should be ignored):

	Proportions
Underweight	0.066
About Right	0.450
Overweight	0.484
Total	1.000

The following bar graph displays the marginal distributions:

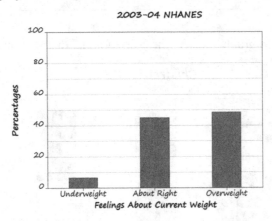

2003-04 NHANES

c. Here is the conditional distribution of weight feelings for each gender:

	Female	Male
Underweight	0.038	0.096
About Right	0.389	0.515
Overweight	0.573	0.389

The following segmented bar graph displays the conditional distributions:

2003-04 NHANES: Feelings About Current Weight

d. The distributions of the two genders do appear to differ here. The men in the sample were 2.5 times more likely than women to feel they are underweight and 1.3 times more likely to feel that their weight is about right. In contrast, women are almost 1.5 times more likely than men to feel that they are overweight.

Exercise 6-15: Preventing Breast Cancer

The analyses for the risk of developing a blood clot in a lung follow:

a. Here is the two-way table:

	Tamoxifen	Raloxifene	Total
Blood Clot	54	35	89
No Blood Clot	9672	9710	19382
Total	9726	9745	19471

b. Here are the conditional proportions of developing a blood clot for each drug:

	Tamoxifen	Raloxifene
Blood Clot	0.006	0.004
No Blood Clot	0.994	0.996

c. The following segmented bar graph displays the conditional proportions:

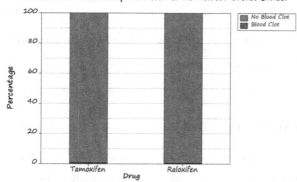

d. The relative risk of developing a blood clot is .006 / .004 ≈ 1.5.

e. You can conclude that for postmenopausal women the risk of developing blood clots in a lung is about 1.5 times greater for taking tamoxifen. You can also conclude that raloxifene is the cause of the decreased risk of lung blood clots because this was a well-designed, randomized, comparative experiment. However, you would need more information about how the women were selected for this study before generalizing to a larger population.

Exercise 6-17: Watching Films

Answers will vary. The following is a representative example:

a. Here is the 2 × 2 table:

	I Saw Movie	I Did Not See Movie	Total
Friend Saw Movie	13	2	15
Friend Did Not See Movie	4	11	15
Total	17	13	30

b. The following segmented bar graph displays the results:

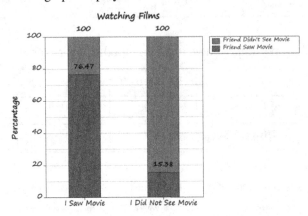

In this case, my friend and I seem to have very similar move-watching habits. If I saw a movie, then there is about a 76% (13/17) chance that my friend also saw it, and if I did not see a particular movie, there is about an 82% (11/13) chance my friend did not see it.

Exercise 6-19: Botox for Back Pain

a. *Randomized* means that the subjects were randomly assigned to two groups – the group that received the Botox injection and the group that received the placebo (saline injection). *Double-blind* means that neither the subjects nor the person administering the injection and/or evaluating their degree of pain relief knew which subjects were in which group.

b. Here is the 2 × 2 table:

	Botox	Saline	Total
Pain Relief	9	2	11
No Pain Relief	6	14	20
Total	15	16	31

c. Here are the conditional proportions for the patients:

	Botox	Saline
Pain Relief	0.6	0.125
No Pain Relief	0.4	0.875

The following segmented bar graph displays these conditional proportions:

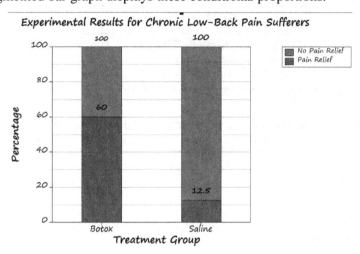

d. The relative risk of pain relief is .6/.125 ≈ 4.8; those receiving Botox were 4.8 times more likely to experience substantial back-pain relief than those receiving saline.

e. Because this was a comparative, randomized experiment, you can conclude that Botox is responsible for increasing the rate of pain relief in patients with chronic low-back pain at the end of eight weeks. However, you should be cautious in generalizing the results from this study to the larger population of back-pain sufferers because you do not know how representative the sample is.

Exercise 6-21: Gender-Stereotypical Toy Advertising

a. Here are the marginal totals:

Boy Shown	Girl Shown
97	86

Traditional "Male" Toy	Traditional "Female" Toy	Neutral Gender Toy
74	26	83

b. The proportion of ads that depict boys with traditional male toys is 59/97 ≈ 0.608.

The proportion of ads that depict boys with traditional female toys is 2/97 ≈ 0.021.

The proportion of ads that depict boys with neutral toys is 36/97 ≈ 0.371.

c. Here is the conditional distribution of toy types for ads showing girls:

	Girl Shown
Traditional "Male" Toy	0.174
Traditional "Female" Toy	0.279
Neutral Gender Toy	0.547

d. The following segmented bar graph displays the conditional distributions:

Toy Advertisements

e. The bar graph indicates that toy advertisers do seem to present pictures of boys with traditional male toys a majority of the time (more than 60%), but the same cannot be said for girls and traditional female toys. Girls tend to be shown with neutral gender toys (about 55% of the time).

Exercise 6-23: Baldness and Heart Disease

a. The proportion of men who identified themselves as having little or no baldness is (251 + 165 + 331 + 221) / (663 + 772) ≈ 0.675.

b. Of those who had heart disease, the proportion of mean who claimed to have some, much or extreme baldness is (195 + 50 + 2) /663 ≈ 0.373.

c. The proportion of men who were in the control group is (331 + 221) / 968 ≈ 0.570.

d. The following segmented bar graph compares the distribution of baldness ratings between subjects with heart disease and those from the control groups:

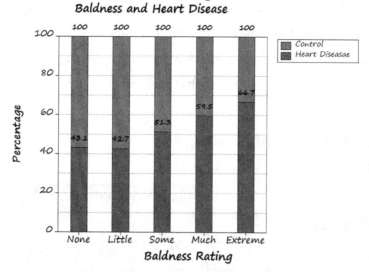

Baldness and Heart Disease

e. In this sample, the more baldness a man has, the more likely he is to suffer from heart disease. Although you would not use this study to conclude a cause-and-effect relationship because this is an observational study, there does appear to be a moderately strong association. You also do not know what population this sample is representative of, so you cannot tell to what population this association might generalize.

f. The risk of heart disease (little or no baldness) is $416/968 \approx 0.430$.

The risk of heart disease (other group) is $247/467 \approx 0.529$.

The relative risk of heart disease is $.529/.430 \approx 1.23$.

Exercise 6-25: Hospital Recovery Rates

a. The proportion of hospital A's patients who survived is 800/1000 or 0.8.

The proportion of hospital B's patients who survived is 900/1000 or 0.9.

Hospital B saved a higher percentage of its patients.

b. The proportion of hospital A's patients in fair condition who survived is $590/600 \approx 0.983$.

The proportion of hospital B's patients in fair condition who survived is $870/900 \approx 0.967$.

Hospital A saved a higher percentage of its patients who were in fair condition.

c. The proportion of hospital A's patients in poor condition who survived is 210/400 or 0.525.

The proportion of hospital B's patients in poor condition who survived is $30/100 \approx 0.300$.

Hospital A saved a higher percentage of its patients who were in poor condition.

d. Hospital A tends to treat a higher proportion of patients who are in poor condition to begin with and whose chances of survival are not very high, regardless of where they seek treatment. Hospital A does a better job with both types of patients, but their overall survival percentage is lower than hospital B's because they treat such a larger rate of patients in poor condition.

e. If you were ill, you should go to hospital A which does a better job of treating both types of patients. No matter how sick you are, you stand a better chance of surviving at hospital A.

Exercise 6-27: Softball Batting Averages

Many answers are possible. Here is one example:

	June	July	Combined
Amy's Hits	80	121	201
Amy's At-bats	100	400	500
Amy's Proportion of Hits	0.8000	0.3025	0.4020
Barb's Hits	319	30	349
Barb's At-bats	400	100	500
Barb's Proportion of Hits	0.7975	0.3000	0.6980

The goal is to have higher batting averages in June (when Barb has most of her at-bats) and lower batting averages in July (when Amy has most of her at-bats).

Exercise 6-29: Politics and Ice Cream

Here is the completed table:

	Chocolate	Vanilla	Strawberry	Total
Democrat	108	96	36	240
Republican	81	72	27	180
Independent	36	32	12	80
Total	225	200	75	500

Notice that the conditional percentages are the same (48%, 36%, and 16%) for each flavor.

Exercise 6-31: Feeling Rushed?

The given graph indicates that the percentage of Americans who always feel rushed is increasing over time. In 1982 less than 25% of Americans claimed to always feel rushed but by 1996 this percentage had grown to 30% and appears to have increased slightly by 2004. During the same time period the percentage of Americans who reported almost never feeling rushed decreased from about 25% in 1982 to about 17% in 2004. The percentage of Americans who claims to sometimes feel rushed appears to have remained essentially unchanged during these years; approximately 50% of the respondents tended to report feeling rushed sometimes in each year.

Exercise 6-33: Government Spending

a. The marginal distribution of opinions about spending on the environment are shown in the table below:

	Proportion
Too Little	0.678
About Right	0.241
Too much	0.081

b. The following bar graph displays this marginal distribution:

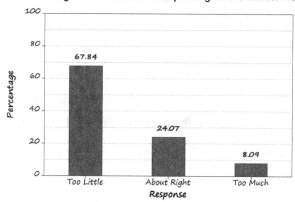

GSS: Feelings About Government Spending on the Environment

This graph reveals that a majority of these subjects believe the federal government spends too little on the environment, and very few respondents believe they government spend too much.

c. The following segmented bar graph displays the conditional distributions of spending opinions for each political viewpoint:

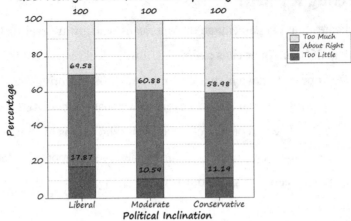

GSS: Feelings About Government Spending on the Environment

d. This graph indicates a clear association between political inclination and feelings about government spending on the environment. The more conservative a person is, the more likely he or she is to feel that the government is spending too much on the environment, and the less likely he or she is to feel the government is spending too little on the environment. If a person classifies herself as a liberal, she is highly likely to believe the government is not spending enough on the environment.

Exercise 6-35: Properties of Independence

a. In order for the variables *gender* and *level of sports participation* to be independent
i. there do *not* have to be 50% and 50% females in the sample. In fact, the percentage of males and females in the sample does not have to be the same.
ii. there does *not* have to be one-third of the sample in each of the three sports participation categories.
iii. the distribution of sports participation levels *does* have to be the same for males and females.
iv. there *does* not have to be at least one person in every sports participation level for independence (though there should be in order for there to be three distinct levels of sports participation).

b. The table below is an example of a numbers of males and females in each of three sports participation categories (none, some, and high).

	Males	Females	Total
None	300	200	500
Some	180	120	300

High	120	80	200
Total	600	400	1000

The marginal distributions of gender and level of sports participation are not split evenly between the categories, however, the conditional distribution of sports participation for each gender are identical as shown in the table and segmented bar graph below:

	Males	Females
None	0.5	0.5
Some	0.3	0.3
High	0.2	0.2

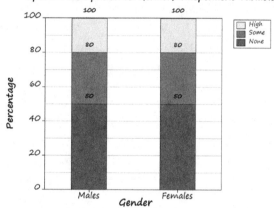

Exercise 6-37: Coffee Consumption

a. Here is the completed table:

	Male	Female	Total
Every Day	(20/70) × 38 ≈ 12	(20/70) × 32 ≈ 10	22
Sometimes	12	10	22
Almost Never	14	12	26
Total	38	32	70

b. The segmented bar graph for the new conditional distributions in which the variables are independent is shown below:

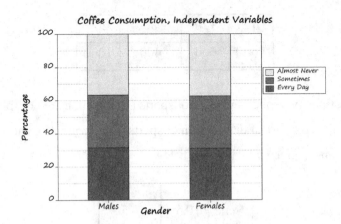

Coffee Consumption, Independent Variables

c. Yes, this graph indicates near independence because the male and female bars show the same breakdowns across the three categories.

Exercise 6-39: In the News

Answers will vary.

Topic 7

Displaying and Describing Distributions

Odd- Numbered Exercise Solutions

Exercise 7-7: Newspaper Data

a. You would expect the shape of this distribution to be skewed to the left. Most of the deceased people would be rather old, but some of them would be unexpectedly young. The following histogram is an example of such a distribution:

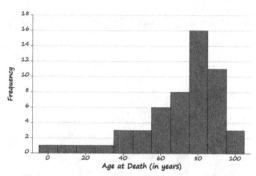

b. You would expect the shape of this distribution to be skewed to the right. Some house prices would be unusually high, and these prices would skew the graph to the right. The following histogram is an example of such a distribution:

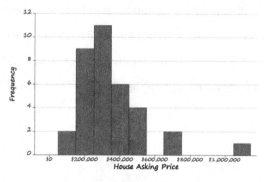

c. You would expect the shape of this distribution to be skewed to the right. Most newspapers will not print stock prices below $1, so they will be truncated on the left. There could also be some stocks with much greater values than the rest. The following histogram is an example of such a distribution:

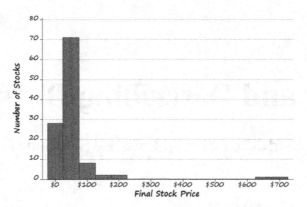

d. You would expect the shape of this distribution to be symmetric. There should be a central cluster, with temperatures tailing off equally to the left and right of center. Another possibility is that you would expect the shape of this distribution to be skewed to the right: Most major cities in the United States would have low temperatures, but there would be a few (L.A., Honolulu, and Miami for example) that would have warmer January temperatures. The following histogram is an example of one possible distribution:

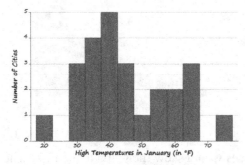

e. You would expect the shape of this distribution to be symmetric. There should be a central cluster, with temperatures tailing off equally to the left and right of center. Another possibility is that you would expect the shape of this distribution to be skewed to the left. Most major cities in the United States would have high temperatures, but there would be a few (Juno, AL, Portland, OR, and Bangor, ME for example) that would have cooler July temperatures. The following histogram is an example of one possible distribution:

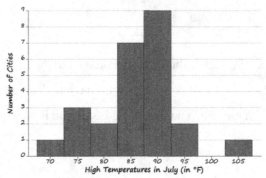

Exercise 7-9: Musical Practice

a. The observational units in this study are the violin students at Berlin's Academy of Music. One variable is *the quality of the violin student (excellent, good, or fair)*, and this variable is categorical. The other variable is *the number of hours spent practicing the violin since first picking it up*. This variable is quantitative.

b. The following histograms reflect the findings in the study conducted at Berlin's Academy of Music (Excellent students average 10,000; good 8,000; and fair 4,000, with very little variability within the distributions):

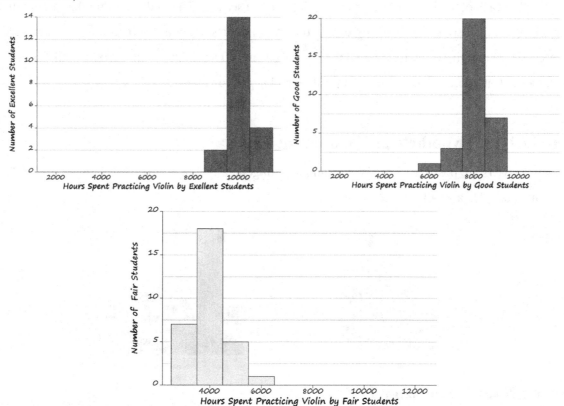

Exercise 7-11: College Football Scores

a. The following stemplot displays the distribution of the margins of victory for the top 25 teams in their first game of the 2009 college football season:

```
-1 | 410
-0 | 41
 0 | 1148
 1 | 01468
 2 | 04
 3 | 14599
 4 | 66
```

$$5 \mid 39$$
leaf unit = one point

b. The distribution of the margins of victory for the top 25 teams for first game of the 2009 college football season is fairly symmetric with a spread from a low of −14 points to a high of 59 points. The center of the distribution is about 18 points. There were five games in which the margins were negative, indicating that top 25 teams were playing each other (except in the case of Florida State's loss to unranked Miami).

c. You would expect the margins to be much smaller, with many more negative values as the top 25 teams play each other and tougher opponents later in the season. This indicates that the mean will be lower. It would also be reasonable to expect the spread to be smaller as the teams are more closely matched (the margins of victory would not be as extreme). The shape would probably still be symmetric (but more balanced between positive and negative values).

Exercise 7-13: Hypothetical Commuting Times

Many answers are possible. Here is an example with generally longer times for A (more above 24 minutes) but also a smaller range of values:

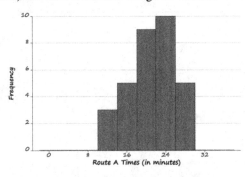 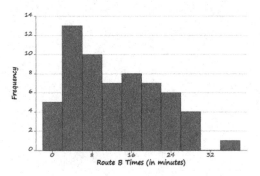

Exercise 7-15: Memorizing Letters

Answers will vary by class. Here is a representative example.

a.

JFK Group		JFKC Group
	0	2333
9998666	0	56666899
2	1	034444455578
98885555	1	04
41111111	2	
77	2	

leaf unit = one letter

b. The stemplots indicate that the JFK group was generally able to memorize more letters than the JFKC group. The center of the JFK group is substantially higher than that of the JFKC group (about 17 letters versus 11 letters).

Exercise 7-17: Hypothetical Manufacturing Processes

The center of the distribution of process A is about 11.5 cm and this distribution does not have a great deal of variability in the rod diameters. In contrast, the rod diameters in process B display a great deal of variability, although their center is right on target at 12 cm. Process C is centered at about 11.7 cm and displays a moderate amount of variability (more than process A). Process D is best as is because its center is right on target (12 cm) at it displays the least variability in rod diameters (so it is also the most stable). Process B is the least stable because it displays the most variability, and process A produces rods with diameters that are generally farthest from the target value.

Exercise 7-19: Jurassic Park Dinosaur Heights

a. Distribution 1 is roughly symmetric with a single peak. Distribution 2 has greater variability with three distinct peaks.

b. Distribution 2, with its three peaks, is what you would expect from a controlled population that had been introduced in three separate batches, whereas Distribution 1 is more consistent with the unimodal shape you would expect from a normal biological population.

c. There are no outliers, and it is almost perfectly symmetric. A distribution occurring in nature would most likely not be so "regular" or perfectly shaped.

Exercise 7-21: Exam Scores

a. Three students received a score of 77 on the exam.

b. Fifteen students scored 90 or greater. This proportion is 15/62, or 0.242.

c. Ten students scored less than 70. This proportion is 10/62, or 0.161.

d. The score that appeared most often is 90.

e. The two values that no one obtained are 78 and 86.

Exercise 7-23: Blood Pressures

The following histogram displays systolic blood pressure measurements:

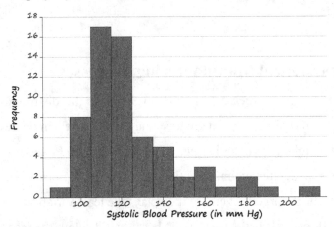

These 64 systolic blood pressures are strongly skewed to the right, ranging from about 90 mmHg to about 185 mmHg with a high outlier at 208 mmHg. There is a peak in the distribution near the typical systolic pressure of about 120 mmHg. The reading of 208 mmHg could be an outlier, as the next highest systolic pressure is 186 mmHg.

The following histogram displays the diastolic blood pressure measurements:

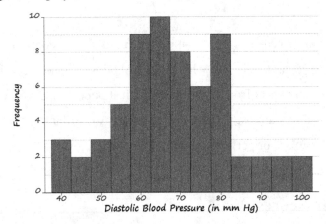

The following stacked dotplot compares systolic and diastolic blood pressure measurements:

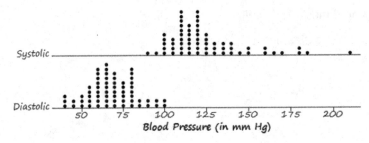

The diastolic blood pressures are much more symmetrically distributed, from a low of about 38 mmHg to a high of 100 mmHg. These pressures appear to be particularly heavily concentrated between 55 and 80 mmHg, with a typical pressure being about 67 mmHg. There are no obvious outliers in this distribution. Both distributions show an interesting granularity in that the pressures are all even values (although this is not clear from the histogram).

Exercise 7-25: British Monarchs' Reigns

a. No, this is not a legitimate histogram of the lengths of reign for the British monarchs. The variable (*lengths of reign*) should be on the horizontal axis (not the vertical axis), and the vertical axis should display the frequencies.

b. No, the distribution is not symmetric. Recall the (correct) stemplot from Exercise 7-24, which showed that the distribution is skewed to the right, with a minimum of 0 and a maximum of 63 years.

Exercise 7-27: Cal State Campus Enrollments

The following histogram displays the distribution of the total number of students enrolled in each campuses of the California State University system in the fall of 2009:

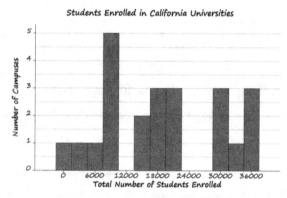

This histogram has 15 bins, each with a width of 2500 students. Using fewer bins hides the clusters and gaps in this distribution. Using more bins creates too much clutter in the plot.

The distribution of the total number of students enrolled in the 23 campuses of the California State University system in Fall 2009 if anything, is skewed to the right, with three distinct clusters separated by clear gaps. Three of the campuses had total enrollments of 5000 or fewer students, and another five campuses enrolled between 7500 and 10,000 students. A cluster of eight campuses enrolled between 125,000 and 225,000 students. The final cluster of seven campuses enrolled somewhere between 275,000 and 375,000 students. The typical campus enrollment appears to be about 18,000 students.

Exercise 7-29: Textbook Prices

a. Dotplots would be the appropriate graphs for comparing the amount spent on textbooks by science and humanities majors, because this (*amount spent on textbooks*) is a quantitative variable. A bar graph should only be used to display categorical data.

b. The variable represented on the horizontal axis of these graphs would be *the amount spent on textbooks this term*.

c. Many answers are possible. The following stacked dotplot displays a distribution in which all six science majors all spent more than six humanities majors on textbooks:

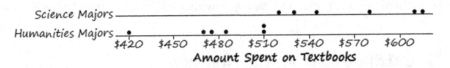

d. Many answers are possible. The following stacked dotplot displays one distribution in which there is a tendency for science majors to spend more than humanities majors on their textbooks:

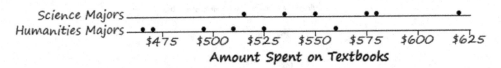

e. Many answers are possible. The following stacked dotplot displays one distribution in the two groups tends to spend similar amounts on average but the amounts show much less consistency among the humanities majors:

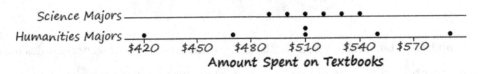

Exercise 7-31: On Your Own

Answers will vary by student and class.

Topic 8

Measures of Center

Odd- Numbered Exercise Solutions

Exercise 8-7: Properties of Mean and Median

a. Yes, it is possible that the mean might not equal any of the values in a dataset. In fact, this is a fairly common occurrence.   For example, the mean weight of the rowers on the 2008 Men's Olympic Rowing team is 197.96 lbs, and this was not the weight of any of the rowers on the team.

b. Yes, it is possible that the median might not equal any of the values in a dataset if there is an even number of values in the dataset. If there are an odd number of values, the median must be the middle number, and therefore must be in the dataset. But if there are an even number of values in the dataset, and the two middle numbers are not the same, the median would be the average of these two middle numbers, and therefore *would not* be in the dataset. For example, the median Republican candidate speaking speed (Activity 8-1) was 188 words per minute, but this value was not one of the speaking speeds of a Republican candidate.

c. No, it is not possible for the mean to be smaller than all the values in a dataset. There must be at least one value that is as large, or larger, than the mean.

d. No, it is not possible for the median to be smaller than all the values in a dataset. Exactly ½ of the values must be as small, or smaller, than the median, and ½ of the values must be as large, or larger.

e. Yes, it is possible for the mean to be larger than only one value in a dataset. If there is one very low outlier, and the remaining values are similar (without too much spread between them), hen all but one of the values could be less than the mean. For example, with the dataset {10, 60, 60, 60, 60, 60 65, 70, 70, 75}, the mean is 59, but only the smallest value (10) is less than the mean.

f. No, it is not possible for the median to be larger than only one value in a dataset because the median has to be in the middle of the ordered data. Exactly ½ of the values must be as small, or smaller, than the median, and ½ of the values must be as large, or larger.

Exercise 8-9: Sampling Words

a. Answers will vary by student expectation.

b. The mean is 4.29 letters. You calculate

$$(7 + 2 \times 29 + 3 \times 29 + 4 \times 59 + 5 \times 34 + \ldots + 9 \times 10 + 10 \times 4 + 11 \times 3) \,/\, 268.$$

The median is 4 letters. You determine the median by finding the 134th word length in the list.

c. You are calculating *parameters* because this is the entire population of words in the Gettysburg Address.

d. The mean cannot possibly be 24.36 letters because no word has more than 11 letters.

e. This hypothetical student added the frequencies (instead of the lengths) and divided by 11 ($268/11 \approx$ 24.36).

Exercise 8-11: Supreme Court Justices

a. The mean length of service is 14.89 years. The median length of service is 16 years.

b. The proportion of the justices that have served for longer than the mean (14.89 years) is $6/9 \approx 0.667$.

c. If Justice Stevens had served for 54 years, the mean would become 17.11 years. The median would not change.

d. The mean and median would each increase by 5 years (to 19.89 and 21 years, respectively).

Exercise 8-13: Memorizing Letters

Answers will vary. Here is one representative set of answers.

a. For the JFK treatment group, the mean is 16 letters and the median is 18 letters. For the JFKC treatment group, the mean is 10.88 letters and the median is 11.5 letters.

b. These are statistics because they come from a sample.

c. The centers indicate that the JFK group was able to memorize more letters successfully, about six more letters on average, than the JFKC group. Yes, the difference appears to be substantial.

d. The grouping of letters does appear to affect memory performance. Because this was a randomized, comparative experiment, you can conclude that it was the grouping that caused the increase in performance.

Exercise 8-15: Population Growth

a. Using all digits (Minitab): For the western states, the median is 8.6%. For the eastern states, the median is 4.7%

Truncating digits: For the western states, the median is 8.5%. For the eastern states, the median is 4.5%.

These medians indicate that the population in the western states is growing, on average, about 4 percentage points faster than in the eastern states.

b. Because both distributions are somewhat skewed to the right, you should expect the means to be larger than the medians.

c. You should expect the mean to decrease significantly if you remove the high outlier, Nevada, from the analysis because the mean is not resistant to outliers.

Exercise 8-17: Marriage Ages

a. The following side-by-side stemplot displays the distribution of ages between husbands and wives:

Wives		Husbands
6	1	9
87765443322	2	3355556699
966320	3	011458
754	4	
0	5	144
0	6	02
3	7	1

leaf unit = one year

b. Both distributions are skewed to the right with most wives and husbands getting married young (in their 20s and 30s). There are two distinct clusters of ages for the husbands, one from 19 to 38 years, and the other from 51 to 71 years, with an overall center in the mid-30s. The wives varied in age from 16 to 71 and were centered a bit younger in their early to mid-30s.

c. For husbands, the mean is 35.71 years and the median is 30.5 years. For wives, the mean is 33.83 years and the median is 29 years.

d. The median of the age differences is 1.0 years

e. No, these two values are not equal. ($1.0 \neq 1.5$)

f. Yes, the mean of the differences is equal to the difference in mean ages (both are 1.875 years).

Exercise 8-19: February Temperatures

a. For Lincoln, the mean is 43.96°F and the median is 42.5°F. For San Luis Obispo, the mean is 67.75°F and the median is 67°F. For Sedona, the mean is 59.61°F and the median is 62°F.

b. For Lincoln, the mean is 6.65°C and the median is 5.83°C. For San Luis Obispo, the mean is 19.86°C and the median is 19.44°C. For Sedona, the mean is 15.34°C and the median is 16.67°C.

c. You calculate Celsius mean/median = (Fahrenheit mean/median − 32) × 5/9.

Exercise 8-21: Tennis Simulations

a. For the standard scoring system, the mean is 6.81 points and the median is 6 points. For the no-ad scoring system the mean is 5.84 points and the median is 6 points. For the handicap scoring system, the mean is 4.79 points and the median is 5 points.

b. The standard system is skewed right, so its mean is greater than its median and is almost a full point greater than the mean of the symmetric no-ad system. Both of these systems have centers more than a full point higher than the handicap system.

Exercise 8-23: Hypothetical Exam Scores

a. No, you do not have enough information to calculate the mean score for the two sections combined. There may be more students in one section than the other.

b. The overall mean must be between 60 and 90.

c. You need to know how many students there are in each section.

d. The overall mean exam score is $[(20 \times 60) + (30 \times 90)]/50$, or 78. The overall mean is closer to 90 because there are more students in Section 2 than in Section 1.

e. Suppose there were 30 students in Section 1 and 2 students in Section 2. Then the overall mean exam score is $[(30 \times 60) + (2 \times 90)] / 32 \approx 61.875$.

f. Yes, you can determine the overall mean if you know the same number of students are in each section. If there are n students in each section, then the overall mean is $[(n \times 60) + (n \times 90)] / 2n$ or 75.

g. If a student's score was greater than the mean of Section 1 (60) but less than the mean of Section 2 (90), then when that student transferred from Section 1 to Section 2, the mean would be lower in Section 1 and also lower in Section 2.

Exercise 8-25: Sports Averages

a. The observational units are a football player, baseball games for a specific team, golfer, hockey team, basketball games for a specific team, and a tennis player. (This assumes the averages are calculated across an individual's (or team's) performance. Alternatively, these could be averages across players or teams.)

b. *Average yards per carry in football*: {3, 4, -2, 0, 14, 7, 2, 5, -4, 3}. The average is 3.2 yards per carry.

Average runs per game in baseball (Atlanta Braves games 5/12/10 – 5/22/10): {9, 6, 1, 13, 2, 3, 5, 10, 7, 4}. The average is 6.0 runs per game.

Average driving distance in golf: {325, 301, 294, 276, 311, 330, 295, 288, 277, 297}. The average is 299.4 yards per drive.

Average goals scored by opponents in hockey (Philadelphia Flyers 2010 Eastern Conference Semifinals): {5, 3, 4, 4, 0, 1, 3}. The average is 2.86 goals per game.

Average points per game in basketball: (Virginia Tech, last 10 games of the regular 2010 season) {74, 70, 72, 61, 87, 55, 60, 100, 71, 88}. The average is 73.8 points per game.

Average speed of a serve in tennis: {120, 118, 115, 95, 121, 119, 105, 120, 107, 115}. The average is 113.5 mph.

c. Answers and justifications will vary but should focus on the expected shape and/or presence of outliers in the distribution.

Average yards per carry in football: If skewed to the right, you would expect the median to be less than the mean.

Average runs per game in baseball: If skewed to the right, you would expect the median to be less than the mean.

Average driving distance in golf: If skewed to the right, you would expect the median to be less than the mean.

Average goals scores by opponent in hockey: If skewed to the left, you would expect the median to be greater than the mean.

Average points per game: If symmetric, you would expect the median to be the same as the mean.

Average speed of a serve in tennis: If skewed to the left , you would expect the median to be greater than the mean.

Exercise 8-27: Readability of Cancer Pamphlets

Here are graphical displays of both distributions:

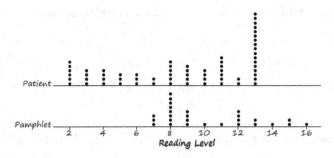

The dotplots are much more useful than only looking at the medians for comparing the two distributions. From the dotplots, you can easily see that the readability levels of the pamphlets are not well-matched to the patients' reading levels. Many patients are not able to read any of the pamphlets. Most patients won't be able to read a large fraction of the pamphlets.

Exercise 8-29: Sleeping Times

a. The observational units are the students in three sections of statistics. The explanatory variable is *section of the statistics course*. This variable is categorical. The response variable is the *student's sleeping time in hours*. This variable is quantitative.

b. The centers are not all similar. The center for section 1 is noticeably smaller than the center for section 2, which is less than the center for section 3.

c. For section 1, the mean sleeping times is $106.25/17 \approx 6.25$ hours.

d. Because there are 17 sleep times in section 1, the median would be the 9th observation, which is 6.25 hours.

e. The mean for section 2 is 7.000 hours and the mean for section 3 is 7.523 hours.
Explanation: The mean for section 2 should be less than the mean for section 3 because there are many observations at 8 or more hours in section 3, but not in section 2, and the sections behave similarly for 6 hours of sleep time or less.

f. The median sleeping time for section 2 is 7 hours and for section 3, the median is 7.5 hours.

g. The mode sleeping time for section 3 is 8 hours. The mode sleeping time for section 2 is 7 hours, and section 1 is bimodal with modes of 6 and 7 hours. These modes indicate the value reported most frequently by students in each section.

Exercise 8-31: Bonus or Rebate?

a. The median amount spent by those in the rebate group is $0. You know this because the median would be the average of the 11th and 12th amounts spent, but the first 16 values were all $0.

b. You cannot determine the median amount spent by the bonus group because the median would be the 13th amount spent if the amounts were all ordered. You know that the first 9 amounts were all $0, but you do not know what any of the other amounts are (including the 13th amount).

You would expect the median to be smaller than $22.04 because the distribution of amounts spent must be skewed to the right. There were 9 students who spent $0, and the remaining students spent a total of $25 \times \$22.04 = \$551$, so the remaining 16 students spent about $34 apiece on average.

Exercise 8-33: Anchoring Phenomenon

a. The mean guess for the Chicago group is $1,140,000. The mean guess for the Green Bay group is $314,400. Yes, the mean for the Chicago group is much larger than the mean for the Green Bay group, as the anchoring phenomenon would suggest.

b. The median guess for the Chicago group is $1,500,000. The median guess for the Green Bay group is $200,000. This difference is also in the direction conjectured by the researchers.

Exercise 8-35: Sleep Deprivation

a. This is an experiment because the researchers randomly assigned the subjects to the control (unrestricted sleep) and treatment (sleep deprived) groups.

b. The explanatory variable is *whether the volunteer was randomly assigned to the unrestricted or sleep deprived group*. This is a binary categorical variable. The response variable is the *improvement (in milliseconds) on the visual discrimination task*. This is a quantitative variable.

c. The stacked dotplots below display the improvement scores of the two groups:

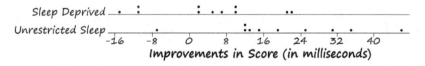

d. For the sleep deprived group, the mean improvement in score was 3.90 milliseconds and the median was 4.5 milliseconds. For the unrestricted sleep group, the mean improvement in score was 19.82 milliseconds and the median was 16.55 milliseconds.

e. Yes, the means and medians seem to support the researchers' hypothesis that sleep deprivation has lingering effects. The mean improvement score for the unrestricted sleep group was 15.92 milliseconds greater than the mean improvement score for the sleep deprivation group. (This difference in mean improvements was actually greater than all but two of the improvements in the sleep deprived group.) The difference in medians was 12.55 milliseconds, in favor of the unrestricted group.

Exercise 8-37: On Your Own

Answers will vary by student.

Topic 9

Measures of Spread

Odd- Numbered Exercise Solutions

Exercise 9-7: February Temperatures

a. Answers will vary by student predictions.

b. For Lincoln, the standard deviation is 15.9° F. For San Luis Obispo, the standard deviation is 9.8° F. For Sedona, the standard deviation is 6.7°F.

Exercise 9-9: Social Acquaintances

Answers will vary by class. Here are some example answers.

The distributions of data collected from both of these classes are very similar. The mean number of acquaintances for the Cal Poly class is 36.1 people, whereas for this class it is 40.8 people. Both classes had minimums below 20 people (6 and 11, respectively) and high outliers above 100 people. The standard deviation for both classes was 25.25 and the IQR for the Cal Poly class is 27 people, whereas for this class it is 29 people. The following stacked dotplots display these results:

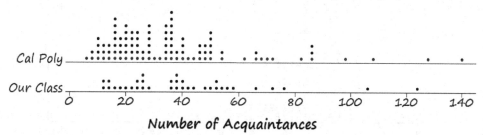

Both distributions appear roughly symmetric apart from the outliers, although the Cal Poly students have a slight skew to the right.

Exercise 9-11: Baby Weights

a. The z-score for Benjamin's weight is $z = (13.9 - 12.5)/1.5$ or 0.93. At age 3 months, Baby Ben was not quite one standard deviation above the average weight.

b. You calculate $0.93 = (x - 17.25)/2$, so $x \approx 19.11$ lbs. If Ben weighs 19.11 lbs. at 6 months, he would again be 0.93 standard deviations above the mean at that age.

Exercise 9-13: Supreme Court Justices

a. Answers will vary by student expectation. Students should consider that moving the entire dataset two values to the right should also move the center by this amount.

b. The mean and median values have increased by 2 years to 16.89 years and 18 years, respectively.

c. Answers will vary by student expectation. Students should realize that adding 2 to each value in the dataset will not change how the data are spread.

d. The IQR and standard deviation did not change; they are 18.5 years and 10.96 years, respectively.

e. Answers will vary by student expectation.

f. The mean is now 29.78 years and the median is now 32 years; the IQR is 37 years and the standard deviation is 21.92 years. All of these values have doubled.

Exercise 9-15: Sampling Words

a. Answers will vary. The following are from one particular running of the applet:

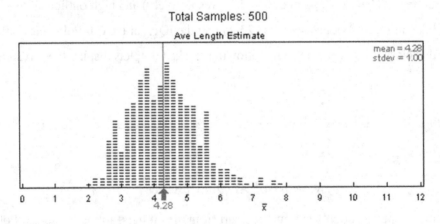

Total Samples: 500

Ave Length Estimate

mean = 4.28
stdev = 1.00

The mean is 4.28 words and the standard deviation is 1.00 words.

b. The applet display is shown below:

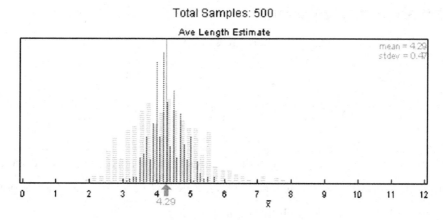

The mean is 4.29 words and the standard deviation is 0.47 words.

c. The standard deviation was roughly cut in half.

d. Yes, for the samples based on a sample of size 20, the empirical rule should hold fairly closely because the sampling distribution is approximately symmetric and mound-shaped.

Exercise 9-17: Baseball Lineups

This stacked dotplot displays the salary distributions for both teams:

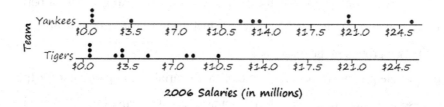

2006 Salaries (in millions)

The Yankees salaries are generally much higher than those of the Tigers and they also exhibit much more variability. The Yankees have a mean salary of $10.96 million and an even greater median of $12.5 million! Their salaries range from a low of $.3 million to a high of $25.6 million and have a standard deviation of $9.45 million and an IQR of more than $20 million. In contrast, the mean Tiger salary is only $4.14 million and the median is a lowly $2.90 million with salaries ranging from $.3 million to a high of only $10.6 million (half of the Yankees make more). The standard deviation of the Tigers' salaries is $3.74 million and their IQR is $7.76 million, reflecting the smaller variability in this distribution.

Exercise 9-19: Memorizing Letters

Answers will vary by class. Those given here are examples.

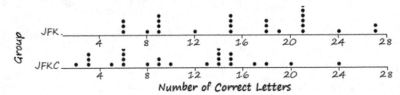

The JFK group showed greater variability in their scores than the JFKC group. The standard deviation for the JFK group was 6.45 letters and their IQR was 12 letters; the JFKC group has a standard deviation of only 5.86 and an IQR of only 9 letters.

Exercise 9-21: Nicotine Lozenge

a. The mean number of cigarettes smoked per day has more variability than the mean age of smoking initiation. You can tell because the standard deviations for number of cigarettes smoked per day are two to three times as large as those for the age of initiation variable.

b. The researchers provide the means and standard deviations so that readers can compare the distributions of the two treatment groups on these baseline characteristics. Showing that these summary statistics are similar adds evidence to the lack of confounding variables between the two treatment groups, which strengthens the causal conclusions from the study if a difference is observed later between the two groups in the response.

c. Yes, the empirical rule probably holds for some of these variables, in particular for the variables *age* and *weight*. These variables are likely to have mound-shaped distributions. It is likely that roughly 68% of these smokers were between the ages of 29 and 53, that 95% of them were between the ages of 17 and 65 and that virtually all of them were between the ages of 5 and 77. Similarly, it is likely that roughly 68% of these smokers weighed between 58.4 and 92.8 kg (129 –205 lbs.), that 95% of them weighed between 41 and 110 kg (9 –242.5 lbs) and that virtually all of them weighed between 24 and 127 kg (53–280 lbs). It is less likely that the age of initiation is symmetric because the mean $- 2 \times SD$ gives an age of 8.3 years (one hopes too young to be realistic). Similarly, the number of cigarettes smoked per day must be truncated at zero, and would not match the empirical rule because the mean $- 3 \times SD < 0$. It also makes sense that extreme chain smokers would skew the distribution to the right.

Exercise 9-23: More Measures

a. The midhinge and midrange are both measures of center because they give the midpoints of the upper and lower quartiles and minimum and maximum values, respectively. This "averaging" should place the results roughly in the middle of the distribution. You would need to look at *differences* between values (e.g., *max – min*) to have a measure of spread.

b. Yes, adding a constant value to all the values in a dataset will change the midhinge and the midrange by that amount. This is further confirmation that these are measures of center because their values change to reflect a shift in the distribution.

c. The midhinge is resistant to outliers because it uses only the upper and lower quartiles in its calculation, and these values are not usually outliers. The midrange is not resistant to outliers because it uses the maximum and minimum values in its calculation, and these are the values that could be outliers.

d. For Botswana, the midrange is (46.5 + 63.5)/2 or 55 years and the midhinge is (48.95 + 62.85)/2 or 55.9 years. For Papua New Guinea, the midrange is (53.9 + 57.4)/2 or 55.65 years, and the midhinge is (55.35 + 56.85)/2 or 56.1 years.

Exercise 9-25: Guessing Standard Deviations

a. Answers will vary by student guess.

b. For Data A, the mean is 64.454 and the standard deviation is 9.598.

  For Data B, the mean is 202.52 and the standard deviation is 51.88.

  For Data C, the mean is 0.99947 and the standard deviation is 0.04952.

  For Data D, the mean is 5.405 and the standard deviation is 4.714.

Exercise 9-27: Super Quarterbacks

The mean number of passing yards for all of Peyton Manning's games is 281.3 yards and the standard deviation is 70.8 yards. Therefore, the z-score for his 16th game is $z = (192 - 281.3)/70.8 \approx -1.26$. This is not an unusual z-score because it is within two standard deviations of the mean. The z-score for Manning's last game is $z = (95 - 281.3)/70.8 \approx -2.63$. This is a mild outlier because it is more than 2 (but less than 3) standard deviations below the mean.

Exercise 9-29: Ages of Restaurant Patrons

Answers will vary by student and school, but the ages of people entering a McDonald's restaurant are likely to have a higher standard deviation than the ages of students entering a snack bar on campus. This is because the students on campus are likely to have ages that are generally in the range from 14 – 18 years, or from 18 – 22 years, whereas the ages of people entering a McDonalds could be quite spread out, from infants to the very elderly.

Exercise 9-31: In the News

Answers will vary by student.

Topic 10

More Summary Measures and Graphs

Odd- Numbered Exercise Solutions

Exercise 10-7: Natural Selection

a. The following boxplots compare the distributions of surviving and perished sparrows for all of these variables:

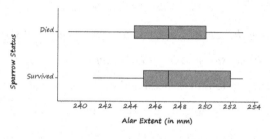

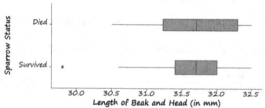

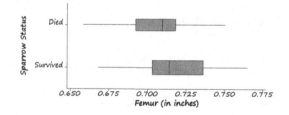

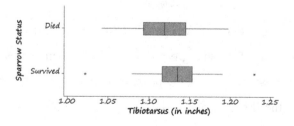

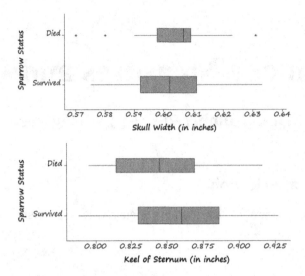

b. The variables *femur bone length*, *keel of sternum*, *humerus bone*, *and skull width* vary considerably. The variables *alar extent, tibiotarsus,* and *length of head and beak* do not vary much.

c. It appears that the alar extent and length of head and beak are virtually the same in the sparrows that did and did not survive the storm. However, the keel of the sternum, the humerus, femur and tibiotarsus bones are all somewhat larger in the sparrows that survived than in the sparrows that died. Only the skull seems to be a little larger in the sparrows that died, although this is hard to say definitively because the spread for the sparrows that died is so much comparatively smaller for this variable than for the sparrows that survived.

Exercise 10-9: Memorizing Letters

Answers will vary by class. One example is given here.

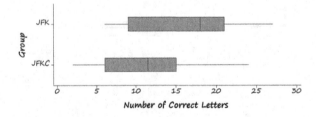

The boxplots indicate that the JFK group was generally more successful at memorizing the letters than the JFKC group, but this group also had a greater variability in their scores. A quarter of the JFK group got 20 letters correct, whereas this same percentage got only 15 letters correct in the JFKC group. More than half of the JFK group did this well.

Exercise 10-11: Sporting Examples

The following boxplots display the distributions of total points between the two sections:

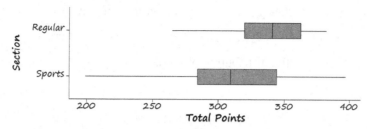

These boxplots show that the students in the regular section tended to earn more points than those in the section that used only sports examples. For the regular section, the minimum is 265 points, with a median of 341 points, a maximum of 262.5 points, and an IQR of 43 points. In contrast, the minimum for the sports-example section was more than 30 points less; 307.25 points, with a median of 309 points and a maximum of 397 points. You see that least 75% of the student scores in the regular section exceeded 50% of the student scores in the sports section. The point spread in the sports section was greater than that of the regular section because their IQR was 60.25 points (versus 43 points).

Exercise 10-13: Social Acquaintances

a. Answers will vary by class. These are representative answers.

Here is a five-number summary of acquaintances: min. = 11, Q_L = 23, median = 37, Q_U = 52, max. = 124

b. $1.5 \times IQR = 1.5 \times (52 - 23) = 1.5 \times 29 = 43.5$. Outliers are outside $[23 - 43.5, 52 + 43.5] = [0, 95.5]$, so there are two high outliers: 105 and 124.

c Here is a modified boxplot of the data:

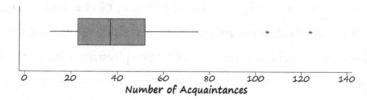

d. The distribution of the number of acquaintances known by students in this class is reasonably symmetric, ranging from a minimum of 11 acquaintances to a maximum of about 75 acquaintances, save for two high outliers at 105 and 124 acquaintances. The center of the distribution is about 37 acquaintances, and the IQR is 29 acquaintances.

Exercise 10-15: Diabetes Diagnoses

a. Here is the five-number summary of these ages: min = 1, Q_L = 39, median = 51, Q_U = 63, max = 88.

b. The IQR is 24 years. You calculate $1.5 \times 24 = 36$, so outliers are any observations outside [3, 99]. Thus, there are four low outliers: 1, 2, 2, and 2. There are no high outliers.

c. The following boxplot displays the distribution of ages:

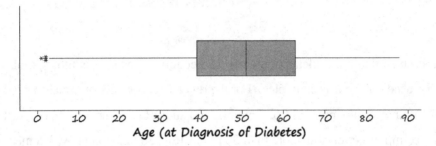

Age (at Diagnosis of Diabetes)

d. The modified boxplot indicates the low outliers which are not obvious on the histogram in Activity 7-5. However, the boxplot does not show the second, small cluster of ages from 2–15 years.

Exercise 10-17: Gender of Physicians

a. The three specialties with the most female physicians are internal medicine (41,658), pediatrics (33,351), and family practice (23,317). The three disciplines with the fewest female physicians are transplantation surgery (8), aerospace medicine (32), and colon/rectal surgery (113).

b. The *number of women* variable does not take into account the total number of physicians in the field.

c. The variable *percentage of that specialty's practitioners who are women* is useful because there are many fields in which only a small number of women practice, but in which only a small number of men practice too. The *relative* number of women in the field gives you a much more informative picture than the *absolute* number.

d. Divide number of women in that specialty by the total number of practitioners (*number of women + number of men*). Then multiply that amount by 100 to convert the proportion into a percentage.

e. The three specialties with the greatest percentage of female physicians are pediatrics (50.25%), medical genetics (46.88%), and child psychiatry (42.32%). The three disciplines with the lowest percentages of female physicians are thoracic surgery (3.01%), urological surgery (3.67%), and orthopedic surgery (3.80%).

f. No, the lists in parts a and e do not agree exactly. This is because the first lists (part a) do not take the total number of physicians in each discipline into consideration.

g. Many answers are possible but medical genetics is an obvious choice. Only 203 women practice in this field, yet this number is 46.9% of specialty's practitioners. This is because so few male doctors (only 230) choose to go into this field; although there are not many female doctors on an absolute scale in this discipline, they make up almost half of the total doctors who specialize in this area.

h. The following dotplot displays the distribution of the percentage of women in these medical specialties:

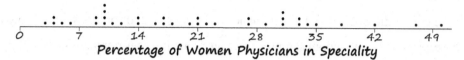

Percentage of Women Physicians in Speciality

The distribution of female physicians' specialties is "mound-shaped", ranging from about 3% (thoracic surgery) to more than 50% (pediatrics), with a center around 20%.

Exercise 10-19: Honda Prices

a. The dotplot for the variable *mileage* corresponds to boxplot I. This plot is strongly skewed to the right with a couple of high outliers.

b. The dotplot for the variable *year of manufacture* corresponds to boxplot III. This plot is skewed to the left.

c. The dotplot for the variable *price* corresponds to boxplot II. This plot has long tails on both sides, although the left tail is much longer than the right.

Exercise 10-21: Rowers' Weights

a. Here is the five-number summary of rowers' weights (in pounds): min. = 121, Q_L = 187.5, median = 205, Q_U = 217, max = 229.

b. You calculate $1.5 \times IQR = 1.5 \times 29.5 = 44.25$, so outliers would be weights outside [187.5 − 44.25, 217 + 44.25] = [143.25, 261.25]. Therefore, there is one low outlier: McElhenney, the coxswain, who weighs 121 pounds.

c. The following boxplot displays the distribution of rowers' weights:

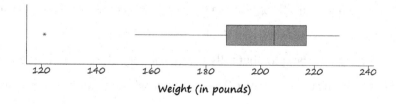

Weight (in pounds)

d. The dotplot reveals the two clusters of weights (lightweight and non-lightweight) that are not shown by the boxplot.

Exercise 10-23: Sampling Words

a. No, the students were not randomly selected from a larger population. The instructor merely used the students who had enrolled in two different sections of his course as his sample. Each student who might take the course did not have an equal chance of being selected for this sample, so you cannot consider these sections to be random samples from any sort of larger population. You might be able to argue that these students are representative of a larger population, but this would be risky.

b. The explanatory variable is the *whether the instructor told the students that the variable of interest was word length in advance*. The response variable was the *sample mean word length (for each student)*.

c. The dotplots indicate that although there is not a substantial difference between the two sections, section 1 tended to choose slightly longer words, on average, than section 2. There was also more variability in the sample mean words length in section 1 than there was in section 2, because four of the students in section 1 chose samples of very long words (their average word lengths were are greater than seven). None of the students in section 1 chose a sample with an average word length less than the population average word length of 4.29 letters per word, whereas one student in section 2 chose such a sample.

Exercise 10-25: Fantasy Draft Times

a. You calculate: IQR = 17 – 6 = 11 seconds and 1.5×IQR = 16.5 seconds. Outliers would be any of the TS draft times that are outside the range [6 – 16.5, 17 + 16.5] or [0, 33.5] seconds. Because the value 107 seconds is larger than 33.5 seconds, it is an outlier.

b. The value 107 seconds would fall in the second 25% of the JW distribution (between Q_L and the median). It would actually be much closer to the lower quartile (102.5 seconds) than it would be to the median (144 seconds).

c. No, there is no player other than TS for which a time of 107 seconds would qualify as an outlier. You can tell by looking at the boxplots shown in the Exercise 10-24 (a) solutions. The value 107 seconds is identified by an * on the TS boxplot, but falls within (or very close to within) all the other boxes, except for that of BK. In the case of player BK, the upper whisker extends to 115 seconds (the highest non-outlier value, which is greater than 107 seconds), so 107 seconds is not an outlier in this distribution either.

Exercise 10-27: Super Quarterbacks

a. Here is the five-number summary of the passing yards for Drew Brees (in yards per game): min = 172, Q_L = 223, median = 298, Q_U = 358, max = 419.

Here is the five-number summary of passing yards for Peyton Manning (in yards per game): min = 95, Q_L = 239.5, median = 302, Q_U = 322.5, max = 379.

b. These five-number summaries indicate that overall, Manning tended to pass slightly fewer yards than Peyton, and that Manning was more consistent the number of yards passed per game than was Peyton. Although Manning has a slightly larger lower quartile and median than Peyton, his minimum, upper quartile and maximum are significantly smaller than Peyton's.

c. For Drew Brees, the IQR is 135. You calculate 1.5 × 135 = 202.5 yards. Outliers would be outside the range [223 – 202.5, 358 + 202.5] = [20.5, 560.5]. To qualify as an outlier, Drew Brees would need to pass fewer than 20 or more than 560 yards in a game.

For Peyton Manning, the IQR is 83. You calculate 1.5 × 83 = 124.5 wpm. Outliers would be outside the range [239.5 – 124.5, 322.5 + 124.5] = [115, 447]. To qualify as an outlier, Peyton Manning would need to pass fewer than 115 or more than 447 yards in a game.

d. Yes, Peyton Manning had a low outlier during the last game of the regular season when he passed only 95 yards. He did not have any high outliers. Drew Brees had no outliers.

Exercise 10-29: Television Viewing Habits

a. The minimum number of hours of television watched by the viewers in this sample is 0. The lower quartile is 1 hour, the median is 2 hours, and the upper quartile is 4 hours. You cannot determine the maximum number of hours watched because you are only told that the largest thirteen values in the sample are "more than 12 hours". You are not told how much more than 12 hours.

b. The IQR is 4 – 1 or 3 hours. You calculate 1.5 × 3 = 4.5 hours. Outliers would be outside the range [1 – 4.5, 4 + .45] = [–3.5, 8.5]. So a viewer would be an outlier if he or she watched fewer than 0 hours or more than 8.5 hours (i.e. 9 or more hours) of television per week. Note that you can tell there are 44 high outliers in this sample. This may seem like a large number of outliers, but it is a relatively small proportion of the sample (0.033).

Exercise 10-31: Mother Ages

a. Using this outlier rule, you calculate $26.882 \pm 2 \times 6.354 = 26.882 \pm 12.708 = [14.174, 39.59]$. Thus the cut-off values would be 14 and 40 years.

b. The cut-off values using this rule (mean $\pm$ 2 $\times$ SD) are less extreme than the cut-offs based on the usual 1.5 $\times$ IQR rule. In other words, using the (mean $\pm$ 2 $\times$ SD) rule will result in more values being identified as outliers than will using the 1.5 $\times$ IQR rule.

c. Yes, you can tell that there is at least one low outlier (the minimum, 13 years), and at least one high outlier (the maximum, 50 years) using the (mean $\pm$ 2 $\times$ SD) rule. You cannot tell if how many additional low or high outliers there may be, if any.

d. Using the (mean $\pm$ 3 $\times$ SD) rule, you calculate $26.882 \pm 3 \times 6.354 = 26.882 \pm 19.062 = [7.82, 65.006]$. Thus the cut-off values using this rule would be 7 and 66 years. These cut-off values are much *more* extreme than the cut-off values for either of the two previous outlier rules, so they will identify far fewer points as outliers. You can tell that there are no outliers in this dataset according to the (mean $\pm$ 3 $\times$ SD) rule because the smallest value (13 years) is larger than 7, and the largest value (50 years) is smaller than 66.

Exercise 10-33: On Your Own

Answers will vary by student.

Unit 3

Randomness in Data

Topic 11

Probability

Odd- Numbered Exercise Solutions

Exercise 11-7: Equally Likely Outcomes

a. These outcomes are equally likely. Because the die is fair, each side should be just as likely to come up as any other side.

b. No, these outcomes are not equally likely. A 2 can only be summed from a roll of snake eyes (1 +1), but a 7 can be summed from rolls of $1 + 6, 2 + 5, 3 + 4, 4 + 3, 5 + 2$, and $6 + 1$. So Pr(sum of 2) = 1/26 whereas Pr(sum of 7) = 6/36 = 1/6.

c. You would assume that a coin flip would have equally likely outcomes.

d. You would assume that spinning a coin would result in equally likely outcomes. (It's actually not!)

e. You would assume this spinning would result in equally likely outcomes

f. No, these outcomes are not equally likely. You might not expect to be just as likely to receive an *F* as an *A*.

g. No, these outcomes are not equally likely. (We hope the probability that California experiences a catastrophic earthquake within the next year is less than ½).

h. No, these outcomes are not equally likely unless you frequent bad restaurants. The probability that your waiter or waitress brings you the meal you ordered is probability greater than ½.

i. No, these outcomes are not equally likely. It seems much more likely that there is not intelligent life on Mars.

j. Answers will vary by personal opinion.

k. Answers will vary by personal opinion.

l. These outcomes are not equally likely. Orange is more likely than the other two colors.

Exercise 11-9: Racquet Spinning

a. The "probability of landing up" refers to the long-run proportion of times that the racquet would land up if it were spun repeatedly under identical conditions. The graph reveals that this proportion seems to be settling in around .45 as the number of spins increases. Many more spins would be needed in order to estimate the probability more accurately.

b. After 10 spins, probability ≈ .3; after 20 spins, probability ≈ .47; after 40 spins, probability ≈ .37; after 100 spins, probability ≈ .45.

c. This graph seems to indicate that landing "down" is a little more likely than landing "up".

d. Answers will vary.

Exercise 11-11: Committee Assignments

a. All possible pairs of officers are listed here: (Alice, Bonnie), (Alice, Carlos), (Alice, Danny), (Alice, Evan), (Alice, Frank), (Bonnie, Carlos), (Bonnie, Danny), (Bonnie, Evan), (Bonnie, Frank), (Carlos, Danny), (Carlos, Evan), (Carlos, Frank), (Danny, Evan), (Danny, Frank), (Evan, Frank).

b. Fifteen pairs are possible. One of these pairs consists of two women.

c. The theoretical probability is $1/15 \approx .06667$. This outcome is uncommon, but not rare.

d. In the long run, two men would be selected as officers 6/15 or 40% of the time. Such an outcome would not be a surprising result using random selection.

e. In the long run, one man and one woman would be selected as officers $8/15 \approx 53.5\%$ of the time. Such an outcome would not be a surprising result using random selection.

f. The most likely outcome would be one officer of each gender.

g. The expected number of men is $0 \times (1/15) + 1 \times (8/15) + 2 \times (6/15) \approx 1.3333$.
This should be reasonably close to the simulated average.

Exercise 11-13: Treatment Groups

Answers will vary. Here is one representative set of answers.

a. Pick any row of the Table of Random Digits. Let even digits represent the women, and odd digits represent the men. Read two digits; this will make up the first group, the second group will be the two remaining people. Repeat 100 times.

b. Suppose the men are Steve and Greg, and the women are Cathy and Laura. Because these are treatment groups, group 1 is different from group 2, thus (Cathy, Laura) / (Steve, Greg) ≠ (Steve, Greg)/ (Cathy, Laura). The sample space is {(Cathy, Laura)/(Steve, Greg), (Cathy, Steve)/(Laura, Greg), (Cathy, Greg)/ Laura, Steve), (Steve, Greg)/(Cathy, Laura), (Laura, Greg)/(Cathy, Steve), (Laura, Steve)/ (Cathy, Greg)}.

Pr(2 *women in one group*, 2 *men in the other*) = 2/6 = 1/3 because Pr(2*W*, 2*M*) = 1/6 and Pr(2*M*, 2*W*)

= 1/6.

Pr(*one of each gender in each group*) = 4/6 = 2/3

Exercise 11-15: Simulating the World Series

a. Answers will vary by student expectation.

b. Answers will vary. Here is one representative set of answers:

Beginning with line 38, the Shorthairs win in 3 games.

c. In 50 simulated series, the Shorthairs won 37 times. This is .74, which is greater than .7.

d. Answers will vary by student expectation.

e. In 50 best-of-seven series, the Shorthairs won 43 times; this is .86.

f. The longer series gives the greater advantage to the better team. With the larger sample size, the proportion of games won by a team is likely to be closer to its probability of winning. Thus, the team with a winning probability greater than ½ is more likely to win more than half of the games in a longer series. In other words, unusual events such as upsets are more likely with smaller sample sizes.

Exercise 11-17: Dice-Generated Ice Cream Prices

a. Pr(*price* = 32¢) = 2/36 ≈ .056; a roll of 3,2 or 2,3 will give a price of 32¢.

b. Pr(*price* = 33¢) = 1/36 ≈ .0278; only a roll of 3,3 will give a price of 33¢.

c. Pr(*price* = 34¢) = 0; if a 3 and 4 are rolled, the price would be 43¢.

d. Pr(*price* < 40¢) = Pr(*rolls of* 11, 12, 21, 13, 22, 23, 31, 32, or 33) = 9/36 = .25.

e. Pr(*price* > 50¢) = Pr(*rolls of* 15, 16, 25, 26, 35, 36, 45, 46, 51, 52, 53, 54, 55, 56, 61, 62, 63, 64, 65, or

66) = 20/36 ≈ .556.

Note: Pr(40 ≤ *price* ≤ 50) = Pr(*rolls of* 14, 24, 34, 41, 42, 43, 44) = 7/36.

f. The sample space is 11¢, 21¢ (including a roll of 12), 31¢ (including a roll of 13), 41¢ (and 14), 51¢ (and 15), 61¢ (16), 22¢, 32¢, (23), 42¢ (24), 52v (25), 62¢ (26), 33¢, 43v (34), 53¢ (35), 63¢ (36), 44¢, 54¢ (45), 64¢ (36), 55, 65¢ (56), 66¢.

The expected value = .11(1/26) + .21(2/36) + . . . + .66 (1/36) = 47.25¢.

g. If you were to repeat this "experiment" over and over again, in the long run, the average price of an ice-cream cone computed in this manner would be 47.25¢.

Exercise 11-19: Hospital Births

Answers will vary. Here is one representative set of answers:

a. The following histograms display the distributions:

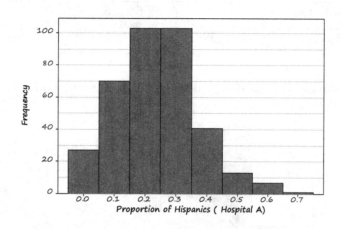

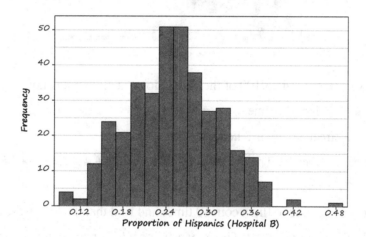

b. Hospital A has more days on which more than 40% of the babies born are Hispanic.

c. Hospital B has more days on which between 15% and 35% of the babies born are of the babies born are Hispanic.

d. Hospital B has more days on which less than 40% of the babies born are Hispanic.

e. Because the sample size (50) for Hospital B is greater than for Hospital A (10), you expect the distribution for Hospital B to be more concentrated around its mean (0.25). So, Hospital B should have a greater percentage of Hispanic births between 15% and 35% and a smaller percentage away from the mean – greater than 40%.

Exercise 11-21: Runs and "Hot" Streaks

a. Answers will vary by guess.

b. Answers will vary. Here is one representative set of answers.

Run Length	2	3	4	5	6	7	8	9
Tally	13	39	25	14	5	2	1	1

c. The following histogram displays the distribution for this set of simulations:

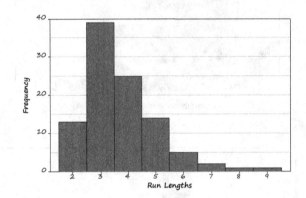

d. A streak of 5 or more occurred in 23/100 of the simulations. This event is not very surprising as it happened almost one-fourth of the time.

e. The most common hot streak length is 3 heads.

f. The median run length is 3 heads. The mean is $(2 \times 13 + 3 \times 39 + 4 \times 25 + 5 \times 14 + 6 \times 5 + 7 \times 2 + 8 \times 1 + 9 \times 1)/100 = 3.74$ heads.

g. Yes, it would be very surprising to flip a coin 10 times and find that the longest streak is 1. In this simulation, a streak of 1 never occurred. The only way it could happen is for the ten flips to alternate back and forth between heads and tails on every toss.

Exercise 11-23: Solitaire

a. The proportion is 74/444 or .167.

b. Answers will vary. Here is one representative set of answers.

c. The following dotplot gives values for this set of results:

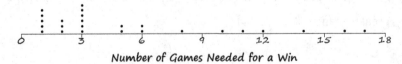

Number of Games Needed for a Win

d. The mean is 5.56 games. The median is 3.0 games.

e. Based on this simulation, this author can expect to play between 5 and 6 games before winning for the first time.

f. Based on this simulation, this author will have to play 3 games in order to have a 50% chance of winning at least once.

g. This author will have to play about 3 games fewer on average before her first win.

Exercise 11-25: Game Show Prizes

a. The fact that the expected value (mean) of the prize amounts is $131,478 does not imply that the most probable outcome is a win of $131,478. In fact, a contestant cannot win this amount because it is not one of the 26 prizes. Each of the 26 prize amounts are assumed to be equally likely, so each of the 26 prizes is as probable as any of the others.

b. No, a contestant does not have a 50% chance of winning more than $131,478 and a 50% chance of winning less than this amount. If all 26 prizes are equally likely, the probability of winning more than $131,478 would be 6/26 because six of the prizes are for amounts greater than this expected value. The probability of winning less than $131,478 would be $20/26 = 10/13 \approx .769$.

c. If this game show were played many hundreds (or thousands) of times, in the long-run, the average amount won by all of the contestants would be $131,478 per game.

Exercise 11-27: Interrupted Game

a. If the game had been able to continue beyond the first six tosses (4 heads and 2 tails), the minimum number of tosses necessary for a winner to be declared would be one (if the next toss were a head, then Hilton would win with 5 heads). The maximum number of tosses that could be played would be three (if the next two tosses were tails, then the third toss would determine the winner regardless of how the coin landed).

b. The sample space for the remaining coin tosses would be {H, TH, TTH, TTT}.

c. No, it is not reasonable to assume the outcomes in this sample space are equally likely because the outcomes do not all consist of the same number of coin tosses. For example, $Pr(H) = ½$, but $Pr(TH) = ½ \times ½ = ¼$.

Exercise 11-29: Interrupted Game

a. The table below shows the number of tosses necessary to complete the game for each outcome in this sample space:

Outcome	HHH	HHT	HTH	HTT	THH	THT	TTH	TTT
Number of Tosses	1	1	1	1	2	2	3	3

b. The table below gives the probability distribution for the number of tosses required to complete the game:

Number of Tosses	1	2	3
Probability	1/2	1/4	1/8

c. The expected value of the number of tosses required to complete the game is

$$1\left(\frac{4}{8}\right) + 2\left(\frac{2}{8}\right) + 3\left(\frac{1}{8}\right) = \frac{11}{8} = 1.375.$$

d. The expected value 1.375 games means that if you were to play finish this interrupted game (after 6 tosses that resulted in 4 heads and 2 tails) many, many times, in the long-run the average number of additional tosses necessary to determine a winner would be 1.375 tosses per interrupted game.

Topic 12

Normal Distributions

Odd- Numbered Exercise Solutions

Exercise 12-5: Normal Curves

a. mean $\approx$ 50, standard deviation $\approx$ 5

b. mean $\approx$ 1100, standard deviation $\approx$ 300

c. mean $\approx$ -20, standard deviation $\approx$ 40

d. mean $\approx$ 225, standard deviation $\approx$ 75

e. It should be clear that the areas of regions I and II are the same and each appears to be about 10% of the total area under the curve. That means the area of region III is approximately 100% – (2 × 10%) or 80% of the total area.

f. Area III appears to be about 10% of the total area. Area II seems to be roughly 30% – 10% or 20% of the total area. This means area I is 100% – 30% = 70% of the total area under the curve.

Exercise 12-7: Professor's Grades

a. The following sketch shows both teachers' grade distributions:

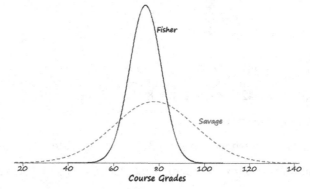

b. Savage gives the higher proportion of As as more than 25% of his grades are As.

For Fisher you calculate Pr(*grade* > 90) = Pr(Z > 2.29) = .0111.

For Savage, you calculate Pr(*grade* > 90) $\approx$ Pr(Z > 0.67) = .2525.

c. Savage also gives a higher proportion of Fs as almost 16% of his grades are Fs.

For Fisher you calculate $\Pr(grade < 60) = \Pr(Z < -2.00) = .0228$.

For Savage you calculate $\Pr(grade < 60) = \Pr(Z < -1.00) = .1587$

d. You have $z = 1.28$. This means you need to solve $(grade - 69)/9 = 1.28$; $grade \approx 80.52$. Therefore you would need above 80.52 in order to earn an A in Professor DeGroot's class.

Exercise 12-9: IQ Scores

a. Here is a sketch of the distribution:

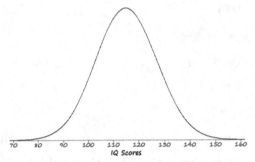

b. Here is a sketch of the distribution with the requested shaded area. Estimates of the proportion of students have an IQ less than 100 will vary by student guess.

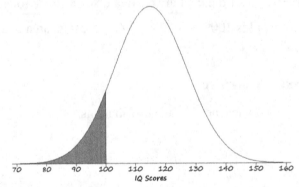

c. $\Pr(IQ\ score < 100) = \Pr(Z < -1.25) = .1057$

d. $\Pr(110 < IQ\ score < 130) \approx \Pr(-.42 < Z < 1.25) = .8944 - .3372 = .5572$

e. $\Pr(IQ\ score < 75) \approx \Pr(Z < -3.33) = .0004$

f. The top 1% (or bottom 99%) corresponds to $z = 2.33$. You calculate $(IQ\ score - 115)/12 \approx 2.33$, then solving for $IQ\ score$ you find $IQ\ score = 142.96$. A student needs an IQ score greater than 142.96 to be in the top 1%.

Exercise 12-11: Candy Bar Weights

a. You want $\Pr(weight < 2.13) = .001$, so $Z = (2.13 - \text{mean}) / .04 = -3.085$. Thus the mean $\mu = 2.25$ oz.

b. You want $\Pr(weight < 2.13) = .001$, so $Z = (2.13 - 2.2) / \sigma = -3.085$. Thus, the standard deviation $\sigma \approx .0227$ oz.

c. You want $\Pr(weight < 2.13) = .001$, so $Z = (2.13 - 2.15) / \sigma = -3.085$. Thus, the standard deviation $\sigma \approx .00648$ oz.

Exercise 12-13: Weights

a. $\Pr(weight < 150) = \Pr(Z < -0.71) = 23.89\%$ (Table II) or 23.75% (applet)

   $\Pr(weight < 200) = \Pr(Z < 0.71) = 76.11\%$ (Table II) or 76.25% (applet)

   $\Pr(weight < 250) = \Pr(Z < 2.14) = 98.38\%$ (Table II) or 98.39% (applet)

b. $\Pr(weight < 150) = \Pr(Z < 0.33) = 62.93\%$ (Table II) or 63.06% (applet)

   $\Pr(weight < 200) = \Pr(Z < 2.00) = 97.73\%$ (Table II) or 97.72% (applet)

   $\Pr(weight < 250) = \Pr(Z < 3.67) = 99.99\%$

c. The normal model does a reasonable job of predicting these percentages. It tends to under estimate somewhat the percentage of both men and women who weigh less than 150 and 200 lbs., but it is very close with the percentages who weigh less than 250 lbs.

Exercise 12-15: Baby Weights

a. You calculate $z = (13.9 - 12.5)/1.5 = .93$. At three months, Benjamin Chance's weight was about 9/10 of a standard deviation above the average.

b. $\Pr(weight > 13.9) = \Pr(Z > .93) = .1762$ (Table II) or .1753 (applet). You must assume that three-month old American babies' weights are normally distributed.

c. By the empirical rule, to be in the middle 68% of weights, his weight would need to be within one standard deviation of the mean, so within $17.25 \pm 2 = 15.25$ lbs. and 19.25 lbs.

Exercise 12-17: Empirical Rule

a. $\Pr(-1 < Z < 1) = .8413 - .1587 = .6826$

b. $\Pr(-2 < Z < 2) = .9773 - .0228 = .9545$

c. $\Pr(-3 < Z < 3) = .9987 - .0014 = .9973$

d. To find the middle 50%, you need 25% on each side. Looking up areas of .25 and .75 in the Standard Normal Probabilities Table (or using technology), you find $z \doteq \pm -.675$. Then you calculate IQR = .675 − (−.675) = 1.35.

e. The z-scores for outliers are −.675 − 1.5×1.35 = −2.7 and .675+1.5 ×1.35 = 2.7.   Using Table II, Pr(−2.7 < Z < 2.7) = .9965 − .0035 = .9930. So, the probability an observation from the normal distribution will be classified an outlier using the 1.5 × IQR rule is 1− .993 or .007.

Exercise 12-19: Body Temperatures

a. The following histogram and normal probability plot display the data on body temperatures:

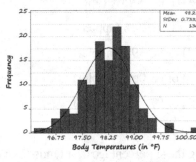 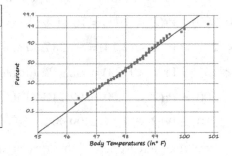

The data seem fairly normally distributed.

b. The following histogram and normal probability plot display body temperature data for men and women:

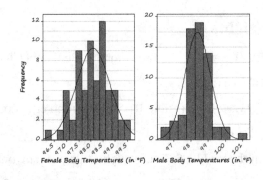

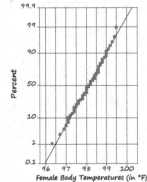 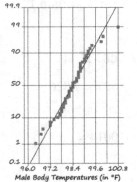

Based on these plots, the female body temperatures seem to more closely follow a normal model. °

Exercise 12-21: Honda Prices

The normal probability plots reveal that prices are approximately normally distributed, but the mileage and year variables are not (skewed to the right and left respectively):

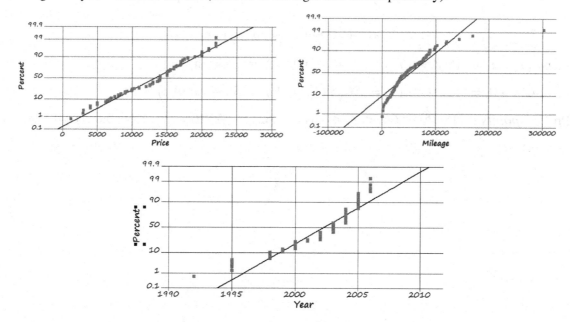

Exercise 12-23: Ultra-High IQ Scores

a. $\Pr(score > 138) = \Pr(Z > 2.53) = .0057$ (Table II) or .0056 (applet)

b. The top 0.1% (or bottom 99.9%) corresponds to $z = 3.09$ (Table II or technology), so then solving for height you find: $3.09 = (score - 100)/15$; $score = 146.35$. A person would have to have an IQ score greater than 146.35 in order to join the Triple Nine Society.

c. Marilyn's z-score is $z = \dfrac{229 - 100}{15} = 8.53$. Therefore Marilyn's IQ is 8.53 standard deviations above the mean.

Exercise 12-25: Football Winning Margins

a. Here is a sketch of the distribution:

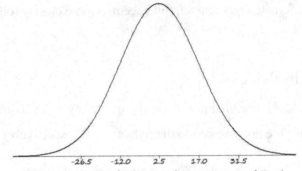

Difference in Number of Points Scored (Home Team–Road Team)

b. Pr(*points* > 14) = Pr(Z > 0.79) = .2148 (Table II) or .2139 (applet)

c. Pr(*points* > 0) = Pr(Z > –0.17) = .5675 (Table II) or .5684 (applet)

d. Pr(–7 < *points* < 7) = Pr(–0.66 < Z < 0.31) = .6217 – .2546 = .3671 (Table II) or .3657 (applet).

Topic 13

Sampling Distributions: Proportions

Odd- Numbered Exercise Solutions

Exercise 13-5: Miscellany

a. Statistic $\hat{p}$

b. Parameter π

c. Statistic $\bar{x}$

d. Parameter μ

e. Parameter σ

f. Statistic s

g. Parameter π (population is all voters)

h. Statistic $\hat{p}$

i. Statistic $\hat{p}$

j. Parameter μ

k. Statistic $\hat{p}$

l. Statistic $\bar{x}$ (population is all American households)

m. Parameter μ

n. Statistic $\hat{p}$

o. Statistic $\bar{x}$

Exercise 13-7: Presidential Approval

a. The standard deviations are given here:

π	0	.2	.4	.5	.6	.8	1
standard deviation	0	.01265	.01549	.01581	.01549	.01265	0

b. The value $\pi = .5$ produces the most variability.

c. The values $\pi = 0$ and $\pi = 1$ produce the least variability.

d. If none (or all) of a population has a particular characteristic, then none (or all) of a sample must have this characteristic as well, leaving no variability in the sample proportion. Similarly, if the population proportion is close to 0 or 1, there is not much "room" for the sample proportion to vary away from the population value. But if exactly half of a population has the characteristic, this should produce the most varied sample proportions.

e. Using a different sample size (500 rather than 1000) would not change the answers to parts b and c (the amount of variability would change, but not the fact that the variability is greatest at $\pi = .5$) because the sample size is a constant in the denominator for the standard deviation for each of these values.

Exercise 13-9: Pet Ownership

a. No, you cannot be certain that the sample proportion of cat households in your sample will be closer to π than your competitor's because of sampling variability, but it is much more likely.

b. Yes, you have a better chance that your competitor of obtaining a sample proportion of cat households that falls within $\pm .05$ of π because you are using a larger sample size.

c. The standard deviations for sample sizes 50 and 200 are given here:

n	50	200
standard deviation	0.061	0.031

The sample size 200 produces the smaller standard deviation – ½ the size of the standard deviation when the sample size is 50 (or two times smaller).

d. Answers will vary. The following is one representative example:

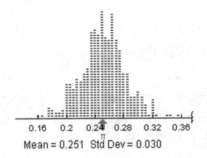

Mean = 0.251 Std Dev = 0.030

The standard deviation is 0.30. This value is extremely close to the value predicted by the Central Limit Theorem, found in part c.

e. Answers will vary. The following is one representative example:

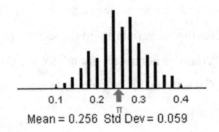

Mean = 0.256 Std Dev = 0.059

The standard deviation in this case is 0.059. This is also fairly close to the standard deviation predicted by the Central Limit Theorem and found in part c.

f. Both distributions are, as expected, approximately normal and centered at .25, but the distribution with samples of size 200 has a much smaller spread than the distribution using samples of size 50. With samples of size 200, the distribution extends from a minimum of only about .16 to a maximum of about .33, and more than 85% of the sample proportions fall within .05 of the mean (.25). In contrast, when the sample size is 50, the sampling distribution extends from 0 to .4 and only about 50% of the sample proportions are within .05 of the mean.

Exercise 13-11: Racquet Spinning

a. Answers will vary. The following is from one representative running of the applet.

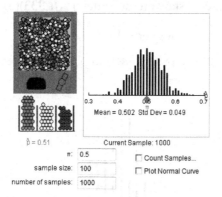

b. The sampling distribution for the proportion of up outcomes is roughly normal with mean .502 and standard deviation .049.

c. The Central Limit Theorem predicts this distribution will be normal with center .5 and standard deviation $\sqrt{(.5)(.5)/100} = .05$. The sampling distribution displayed by the applet simulation is very close to this prediction.

Exercise 13-13: Distinguishing Between Colas

a. If the subject is just guessing, the proportion he or she would get correct in the long run is $\pi = 1/3$.

b. Roll the die 30 times to represent the 30 trials. If you roll a 5 or 6, consider this a success (you successfully identified the odd cola). Otherwise (if you roll a 1, 2, 3 or 4), you failed to identify the odd cola.

c. Simulations will vary by student, but can be performed using the Reese's Pieces applet by using a "*sample size*" of 30, letting the "*number samples*" be 1000, and setting $\pi = .3333$.

d. Below is a histogram obtained from one simulation using the applet:

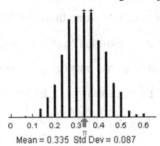

Yes, the shape of this sampling distribution is roughly normal.

e. The mean of the samples in this simulation is .335; the standard deviation is .087. The mean predicted by the CLT is .333 and the predicted standard deviation is $\sqrt{(.333)(.667)/30} \approx .086$. The simulated sampling distribution and CLT values are very close.

f. The applet reports that in 180/1000 or 18% of the samples, the subject guessed correctly 40% or more of the time.

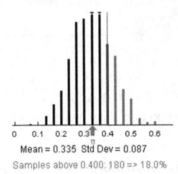

g. Using the Central Limit Theorem, $\Pr(\hat{p} \geq .40) = \Pr(Z \geq (.40 - .3333)/.086) = \Pr(Z \geq 0.76) = .2190$.

Normal Probability Calculator

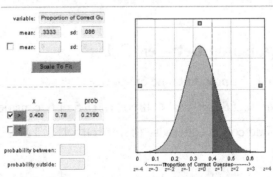

h. The simulation and CLT approximation values are fairly close (.18 vs. .22).

Exercise 13-15: Candy Colors

a. The Central Limit Theorem predicts the sample proportion of orange candies will have a distribution that is approximately normal with a center (mean) of 0.45 and a standard deviation of

$$\sqrt{(.45)(.55)/75} = 0.057.$$

b. Here is sketch of this sampling distribution with the area corresponding to the probability that a sample proportion of orange candies will be less than .4 shaded:

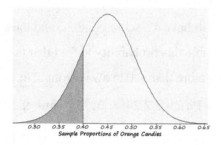

Sample Proportions of Orange Candies

Answers will vary by student guess.

c. $\Pr(\hat{p} < .4) = \Pr\left(Z < \dfrac{.4 - .45}{.057}\right) = \Pr(Z < -0.87) = .1922$ (Table II) or .1902 (applet).

d. You calculate $\Pr(.35 \le \hat{p} \le .55) = \Pr\left(\frac{.35-.45}{.057} \le \hat{p} \le \frac{.55-.45}{.057}\right) = \Pr(-1.75 \le Z \le 1.75) = .9599 - .0401 = .9198$ (Table II) or .9206 (applet).

e. Answers will vary by student. The following are the results of one particular running of the applet:

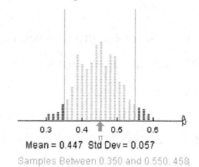

Mean = 0.447   Std Dev = 0.057
Samples Between 0.350 and 0.550: 458

The number of sample proportions that fell within ± .10 of .45 is 458. This is 458/500 or 91.6%. This is very close to the probability found in part d.

Exercise 13-17: Smoking Rates

a. The CLT predicts this distribution will be approximately N(.276, $\sqrt{.276(1-.276)/400} = .02235$).

Here is a sketch of the sampling distribution of the sample proportion of Kentucky smokers:

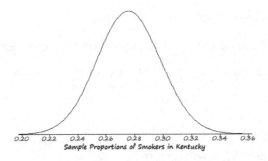

0.20 0.22 0.24 0.26 0.28 0.30 0.32 0.34 0.36
Sample Proportions of Smokers in Kentucky

b. You calculate $\Pr(\hat{p} < .251) = \Pr\left(Z < \frac{.251-.276}{.02235}\right) = \Pr(Z < -1.12) = .1314$. Because the normal distribution is symmetric, .301 will have a z-score of +1.12 and the area to the right of $z = 1.12$ will also be .1314. Thus, you can double this probability to find that the probability of obtaining a sample proportion of Kentucky smokers more than 0.025 away from .276 is $2 \times (.1314) = .2628$.

c. $\Pr(\hat{p} < .226) = \Pr\left(Z < \frac{.226-.276}{.02235}\right) = \Pr(Z < -2.24) = .0125$. Thus, the probability of obtaining a sample proportion of Kentucky smokers more than 0.05 away from the population proportion (.276) is $2 \times (.0125) = .025$.

d. You have no reason to doubt that the state is Kentucky because part b shows that if the state *is* Kentucky, you have a greater than 26% chance of finding a sample result of at least 25% smokers.

e. Now you would have reason to doubt that the state is Kentucky because part c shows that if the state *is* Kentucky, you have less than a 2.5% chance of finding a sample result of at least 22.5% smokers.

Exercise 13-19: Spam

a. The 80% value is a parameter because it is a value that represents the entire population of email messages for the college's computer system. The value, written as the proportion .80, would be represented with the symbol π.

b. In order to apply the CTL, we need to check that the sample is randomly selected. √
    We also need $200(.80) = 160 > 10$ √ and $200(.20) = 40 > 10$ √ .

c. Using the CLT, the sampling distribution of this sample proportion would have a mean of .80 and a standard deviation of $\sqrt{(.80)(.20)/200} = .028284$.

d. $\Pr(\hat{p} \geq .75) = \Pr\left(Z \geq \frac{.75-.80}{.028284}\right) = \Pr(Z \geq -1.77) = .9616$ (Table II) or .9615 (applet).

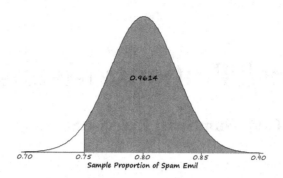

0.9614

Sample Proportion of Spam Emil

e. In order for the CLT to be valid you require both $n(.80) \geq 10$ and $n(1 - .80) \geq 10$. This means you need $n \geq 10/.80 = 12.5$ and $n \geq 10/.20 = 50$. Thus the smallest sample size that would allow you to implement the CLT would be 50.

Topic 14

Sampling Distributions: Means

Odd- Numbered Exercise Solutions

Exercise 14-5: Heart Rates

a. Yes, this distribution appears to be approximately normal, although it is a little skewed to the right.

b. Here is a normal probability plot of the heart rates:

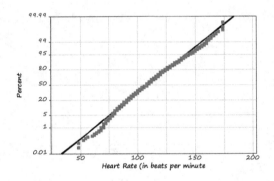

The probability plot indicates that this data is not entirely normally distributed; the values in both the left and right tails are both a little larger than we would expect, indicating some skewness to the right, but the deviations from normality are slight.

c. The following graph displays the distribution of the 1000 sample means with $n = 3$:

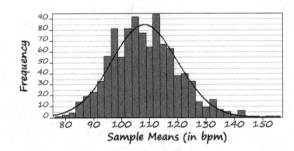

This distribution appears approximately normally distributed (with a slight right skew) with mean 108.25 bpm and standard deviation 11.79 bpm.

d. Yes, the distribution appears roughly normal in spite of the small sample size. This is because the population itself was so close to being normally distributed.

e. The following graph displays the distribution of the 1000 sample means with $n = 10$:

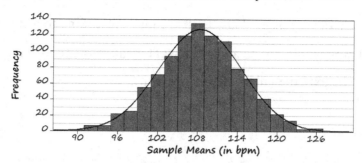

Sample Means (in bpm)

This distribution is approximately normally distributed with mean 108.4 bpm and standard deviation 6.245 bpm. With the larger sample size, the distribution is more normal (less skewed to the right), and of course the spread is substantially smaller.

Exercise 14-7: Car's Fuel Efficiency

a. No, it would not be surprising to obtain 30.4 mpg for one tankful. This value is well within one standard deviation of the mean as shown on a plot of the sampling distribution of the Passat's fuel efficiency:

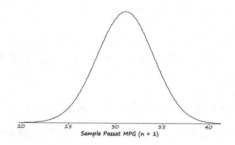

Sample Passat MPG (n = 1)

b. The CLT predicts a sampling distribution that would be approximately normal with mean 31 mpg and standard deviation $3/\sqrt{30} = .5477$ mpg. According to a sketch of this sampling distribution, 30.4 mpg would not be an unusual mean value for a sample of 30 cars.

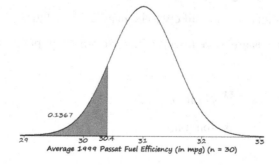

Average 1999 Passat Fuel Efficiency (in mpg) (n = 30)

c. The CLT predicts a sampling distribution that would be (approximately) normal, with mean 31 mpg and standard deviation $3/\sqrt{60} = 0.3873$ mpg. According to a sketch of the sampling distribution, 30.4 mpg would not be a very unusual mean value for a sample of 60 cars to obtain (it is not in the extreme tail of the distribution).

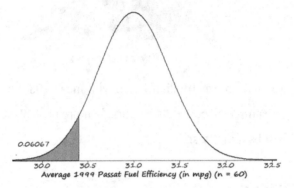

d. Yes, you need to know the population distribution is normal in order to determine this for the case when the sample size is $n = 1$. You do not need to know the shape of the population distribution when $n = 30$ or $n = 60$ as the Central Limit Theorem should still apply regardless of the population shape (as long as it is not too extreme).

Exercise 14-9: Birth Weights

a. The first histogram is from samples of size 10 because it has less variation (a smaller standard deviation) than the second histogram.

b. Sample size 5 is more likely to produce a sample mean birth weight less than 2500 grams.

c. Sample size 5 is more likely to produce a sample mean birth weight less than 3000 grams.

d. Sample size 5 is more likely to produce a sample mean birth weight greater than 3500 grams.

e. Sample size 10 is more likely to produce a sample mean birth weight between 3000 and 3500 grams.

f. The larger the sample size, the smaller the standard deviation of the sample mean, which means the smaller the spread of the sampling distribution of the sample means. This means that with a large sample size, the sample means are more likely to be near the population mean, and with a small sample size, the sample means are more likely to be far from the population mean.

Exercise 14-11: Candy Bar Weights

a. Here is a sketch of the sampling distribution:

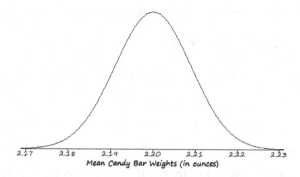

Mean Candy Bar Weights (in ounces)

b. This sampling distribution is normal (shape), and is centered at 2.2 ounces, just as the sampling distribution in the previous exercise was. However, this sampling distribution has a much smaller spread (standard deviation = 0.009).

c. With the smaller value of σ, the probability a same mean weight would be less than 2.15 ounces would decrease significantly because there would be fewer sample means so far below the population mean 2.2 ounces.

d. With this smaller value of σ, the probability a same mean weight would be less than 2.18 ounces would also decrease significantly because there would be fewer sample means this far below the population mean 2.2 ounces.

e. With this smaller value of σ, the sample mean weight that would have a probability of .025 of obtaining a smaller sample mean weight would increase. This is because the needed *z*-score would not change ($z = -1.96$), but the corresponding *sample mean weight* would be calculated using the smaller standard deviation, and so the necessary *sample mean weight* would be closer to the population mean (2.2 ounces).

Exercise 14-13: IQ Scores

It is more likely that a randomly selected resident will have an IQ greater than 120 than a sample of 10 residents will have an average IQ greater than 120. It is much more likely that a randomly selected individual will have an IQ far from the center than that the average IQ of a group will be far from the center. It is really hard to select a group that is "different" from average, but it is not so hard to select a single individual who is different from "average".

Exercise 14-15: Soda Can Volumes

a. $\Pr(soda\ can\ volume > 350) = \Pr\left(Z > \frac{350-358}{6}\right) = \Pr(Z > -1.33) = .9082$ (Table II) or .9088 (applet).

b. The CLT predicts the sampling distribution of the sample mean of a six-pack will be (approximately) $N(358 \text{ ml}, 6/\sqrt{6} = 2.45)$. You calculate $\Pr(\bar{X} > 350) = \Pr\left(Z > \frac{350-358}{2.45}\right) = \Pr(Z > -3.27) = .9995$.

c. The calculation in part b is valid even though the sample size is small because the population is normally distributed.

d. The answer to part b is larger than the answer to part a. This makes sense because the standard deviation of a sample of six cans is smaller than the population standard deviation. This makes the sample mean of six cans more likely to be close to the population mean (358 ml), and means that virtually all samples of six cans will have a sample mean of at least 350 ml.

Exercise 14-17: Birth Weights

a. The z-score is $(2500 - 3300)/507 = -1.40$, so $\Pr(weight < 2500) = \Pr(Z < -1.40) = .0806$ (Table II) or .0802 (applet).

b. Now the standard deviation will be $570/\sqrt{2} = 403.05$ grams. The z-score is $(2500 - 3300)/403.5 = -1.99$, so $\Pr(\bar{X} < 2500) = \Pr(Z < -1.99) = .0233$ Table II) or .0236 (applet).

c. This probability is less than the probability in part a. This makes sense because you are looking at an average; it is harder for any pair of babies to have an average birth weight less than 2500 grams than it is for a single baby to weigh less than this amount.

d. Now the standard deviation will be $570/\sqrt{4} = 285$ grams. The z-score is $(2500 - 3300)/285 = -2.81$, so $\Pr(\bar{X} < 2500) = \Pr(Z < -2.81) = .0025$.

   This probability is much less than the probability in part a. This makes sense because you are looking at an average of four babies; it is harder for four babies to have an average birth weight less than 2500 grams than it is for a single baby to weigh less than this amount.

e. $\Pr(3000 \leq weight \leq 3600) = \Pr\left(\frac{3000-3300}{570} \leq Z \leq \frac{3600-3300}{570}\right) = \Pr(-.53 \leq Z \leq .53) = .4038$ (Table II) or .4013 (applet).

f. Answers will vary by student expectation.

g. Now the standard deviation will be $570/\sqrt{20} = 127.46$ grams. You calculate
   $\Pr(3000 \leq weight \leq 3600) = \Pr\left(\frac{3000-3300}{127.46} \leq Z \leq \frac{3600-3300}{127.46}\right) = \Pr(-2.35 \leq Z \leq 2.35) = .9812$ (Table II) or .9814 (applet).

Exercise 14-19: IQ Scores

a. $\Pr(score > 110) = \Pr\left(Z > \frac{110-105}{12}\right) = \Pr(Z > 0.42) = .3372$ (Table II) or .3385 (applet).

b. $SD = 12/\sqrt{10} = 3.795$.

   $\Pr(\overline{X} > 110) = \Pr\left(Z > \frac{110-105}{3.795}\right) = \Pr(Z > 1.32) = .0934$ (Table II) or .0935 (applet).

c. $SD = 12/\sqrt{40} = 1.897$.

   $\Pr(\overline{X} > 110) = \Pr\left(Z > \frac{110-105}{1.897}\right) = \Pr(Z > 2.63) = .0043$ (Table II) or .00042 (applet).

d. Yes, the calculation in part c would be valid even if the distribution of IQs in the population were skewed because the sample size is large (40 > 30).

Topic 15

Central Limit Theorem and Statistical Inference

Odd- Numbered Exercise Solutions

Exercise 15-5: Miscellany

a. Sample mean

b. Sample proportion

c. Sample mean

d. Sample mean

e. Sample proportion

Exercise 15-7: Christmas Shopping

a. No, it is not valid to use the CLT because the sample size is too small and you do not know that the population is normally distributed.

b. Yes, with a sample of size 500, the CLT predicts the sampling distribution of the sample means would be approximately N($850, $250/$\sqrt{500}$ = $11.18). Here is a sketch of the (approximate) sampling distribution:

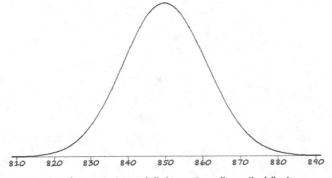

Sample Average Expected Christmas Expenditures (in dollars)

c. $\Pr(\$831.61 < \bar{X} < \$868.39) = \Pr(\frac{831.61-850}{11.18} < Z < \frac{868.39-850}{11.18}) = \Pr(-1.645 < Z < 1.645) = .9495 - .0505 =$ .8990 (Table II) or .8989 (applet).

d. $\Pr(\$828.09 < \bar{X} < \$871.91) = \Pr(\frac{828.09-850}{11.18} < Z < \frac{871.91-850}{11.18}) = \Pr(-1.96 < Z < 1.96) = .9750 - .0250 =$ .9500 (Table II) or .9499 (applet).

e. $\Pr(\$821.20 < \bar{X} < \$878.80) = \Pr(\frac{821.20-850}{11.18} < Z < \frac{878.80-850}{11.18}) = \Pr(-2.58 < Z < 2.58) = .9951 - .0049 =$ .9901 (Table II) or .9900 (applet).

f. First, find the z-scores that mark 80% of the area in the middle of the standard normal curve:

$\Pr(-z* < Z < z*) \approx .8000 \quad \rightarrow \quad \Pr(-1.28 < Z < 1.28) \approx .8000$

As $z = (\bar{x} - 850)/11.18$, and $z = 1.28$, $\bar{x} = (1.28) \times (11.18) + 850 = \$864.31$ Therefore, $k = \$864.31 - \$850 = \$14.13$

g. $\Pr(\$981.61 < \bar{X} < \$1018.39) = \Pr(\frac{981.61-1000}{11.18} < Z < \frac{1018.39-1000}{11.18}) = \Pr(-1.64 < Z < 1.64) = .9495 - .0505 =$ .8990 (Table II) or .8989 (applet). This is exactly what you found in part c. The probability of falling within ± \$18.39 of μ is the same, regardless of what value you use for μ.

Exercise 15-9: Body Temperatures

a. The following graph displays the distribution of body temperatures.

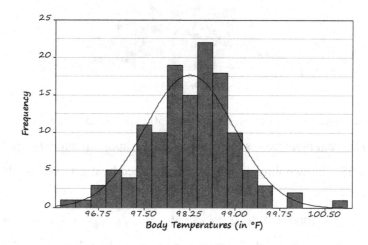

These body temperatures are fairly normally distributed with a couple of high outliers greater than 100°F. The center appears to be around 98.3°F and the values range from roughly 96.5 to 110.5°F.

b. The sample mean is 98.249°F and the standard deviation is 0.733°F.

c. The CLT says the sampling distribution would be N(98.6,.7$/\sqrt{130}$ = 0.061394), so Pr($\bar{X}$ < 98.249)

$= \Pr(Z < \frac{98.249 - 98.6}{0.061394}) \approx \Pr(Z < -5.72) = 0.00$.

d. Yes, the probability found in part c is low enough to provide compelling evidence that the population mean body temperature is not 98.6 degrees. If it were, you would never (probability zero) find a sample of 130 healthy adults with an average body temperature as low as 98.249°F. Because you did find such a sample, you should not believe the population mean temperature is as high as 98.6°F.

Exercise 15-11: Racquet Spinning

a. The value .50 is a parameter because it describes the long-run result of the spinning process.

b. The value .46 is a statistic because it is the result of a sample.

c. Answers will vary. The answers given here are from one particular running of the applet.

The following graph shows the distribution of the 1000 sample proportions:

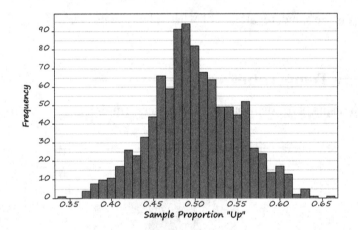

This distribution is approximately normal, with mean .500 and standard deviation 0.0513.

The Central Limit Theorem predicts this distribution will be approximately normal, with mean .5 and standard deviation 0.05. The sampling distribution determined by the applet simulation is very close to this.

d. The applet reports that 250/1000 or 25% of the samples had a sample proportion of at least .54 and 243/1000 or 24.3% of the samples had a sample proportion of .46 or less. Together this is 52.3% of the samples.

e. $\Pr(\hat{p} \leq .46) + \Pr(\hat{p} \geq .54) = \Pr(Z \leq \frac{.46-.5}{.05}) + \Pr(Z \geq \frac{.54-.5}{.05}) = \Pr(Z \leq -0.80) + \Pr(Z \geq 0.80) = .2219 +$

.2219 = .4438. This theoretical approximation is reasonably close to the simulation results of part f.

f. This answer suggests that .46 is not very unlikely to occur by chance alone if the results are 50/50 in the long run. Such an outcome will happen about 45% of the time by chance alone, so it is certainly not rare.

Exercise 15-13: Distinguishing Between Colas

a. If a subject was simply guessing, the CLT predicts the sampling distribution would be roughly $N(.333, \sqrt{.333(1-.333)/30} = .086)$. A sketch of the sampling distribution is shown below.

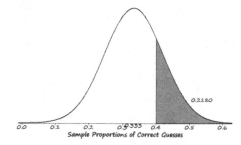

Based on this sketch, you would not be convinced he or she was doing better than guessing would allow in this case. If the subject were simply guessing, he or she would get 40% or more correct about 22% of the time, so this outcome is not all that surprising for a guesser.

b. If a subject were correct 60% of the time in this experiment, based on a sketch of the sampling distribution predicted by the CLT, you *would* be convinced that he or she was doing better than guessing would allow because such an observation is now in the far right tail of the distribution.

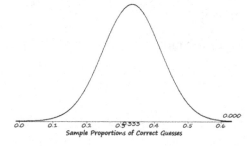

c. Answers will vary. The following is based on one particular running of the Reese's Pieces applet:

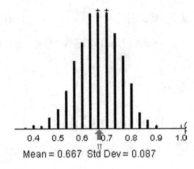

Mean = 0.667 Std Dev = 0.087

d. The rough shapes of the histograms would look like:

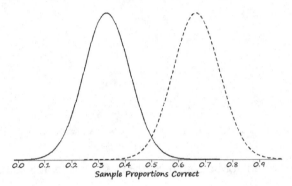

Sample Proportions Correct

Both distributions are approximately normal and have the same spread. However, the have different centers. This is little overlap in the distributions.

e. The applet reports that in 727/1000 or 72.7% of the samples, the subject guessed correctly 60% or more of the time.

Exercise 15-15: Pet Ownership

a. Here is a sketch of this sampling distribution:

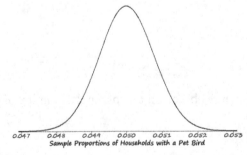

Sample Proportions of Households with a Pet Bird

The CLT says this sampling distribution will be approximately normal, centered at .05, with a standard deviation of $\sqrt{.05(1-.05)/80,000} \approx .000771$.

b. This standard deviation is *much* smaller because you are assuming π is smaller (.05 rather than .333) and farther away from .5.

c. You calculate $z = (.046 - .05)/.1000771 \approx -5.19$. This is a very unusual z-score. $\Pr(Z < -5.19) \approx 0$, so the survey does provide strong evidence that the population proportion who own a pet bird is not 5%. (You observed a sample result that pretty much never happens when $\pi = .05$, so you are convinced $\pi \neq .05$.

Exercise 15-17: Pursuit of Happiness

a. You calculate $z = (.4 - .8)/.007286 = 5.49$.

b. Yes, this z-score is extreme enough to cast doubt on the assertion that 80% of the population felt happy. $\Pr(Z > 5.49) \approx 0$, so if π really is .80, you would never expect to see such a sample result – yet you did see this sample result, so you have strong evidence that π is not .80.

c. $\pi = .82; z = \dfrac{.84 - .82}{\sqrt{\dfrac{(.82)(.18)}{3014}}} = 2.858$ $\qquad$ $\pi = .83; z = \dfrac{.84 - .83}{\sqrt{\dfrac{(.83)(.17)}{3014}}} = 1.462$

$\pi = .84; z = \dfrac{.84 - .84}{\sqrt{\dfrac{(.84)(.16)}{3014}}} = 0.00$ $\qquad$ $\pi = .85; z = \dfrac{.84 - .85}{\sqrt{\dfrac{(.85)(.15)}{3014}}} = -1.538$

$\pi = .86; z = \dfrac{.84 - .86}{\sqrt{\dfrac{(.86)(.14)}{3014}}} = -3.164$ $\qquad$ $\pi = .87; z = \dfrac{.84 - .87}{\sqrt{\dfrac{(.87)(.13)}{3014}}} = -4.897$

$\pi = .88; z = \dfrac{.84 - .88}{\sqrt{\dfrac{(.88)(.12)}{3014}}} = -6.758$

Plausible values of the population proportion include .83 – .85 because these values lie within two standard deviations of the observed sample proportion.

Exercise 15-19: Candy Bar Weights

a. If samples of size $n = 5$ are taken repeatedly, the CLT predicts the sampling distribution of the sample mean will be $N(2.2, \ 0.04/\sqrt{5} = .0179)$. Here is a sketch of this sampling distribution:

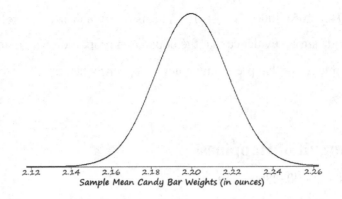

Sample Mean Candy Bar Weights (in ounces)

b. Yes, it is possible to get a sample mean weight as low as 2.15 ounces even if $\mu = 2.2$ ounces is true.

c. Yes, it is fairly unlikely to get a sample mean weight this low if the manufacturer's claim that $\mu = 2.2$

is true. $\Pr\left(\overline{X} < 2.15\right) = \Pr\left(Z < \frac{2.15-2.2}{0.0179}\right) = \Pr(Z < -2.79) = .0026$. If the manufacturer's claim that $\mu =$

2.2 ounces is true, the probability of obtaining a random sample of five candy bars with a mean

weight as small as 2.15 ounces is only .26%.

d. Because it is so unlikely to find a sample of five bars with a mean weight of 2.15 ounces if μ really is

2.2 ounces, this sample provides very strong evidence that μ is actually smaller than 2.2 ounces.

e. You calculate $\Pr\left(\overline{X} < 2.18\right) = \Pr\left(Z < \frac{2.18-2.2}{0.0179}\right) = \Pr(Z < -1.12) = .1314$ (Table II) or .1319 (applet).

This does not provide much evidence against the manufacturer's claim that $\mu = 2.2$ ounces because

such a sample result would occur about 13% of the time when $\mu = 2.2$ ounces.

f. Using the empirical rule, sample means that are more than 2 standard deviations away from the

(assumed) population mean of 2.2 ounces would provide fairly strong evidence against the

manufacturer's claim. Thus sample means that are outside the range $2.2 \pm 2 \times 0.0179 = [2.1642,$

2.2358] would provide fairly strong evidence against the manufacturer's claim.

Unit 4

Inference from Data: Principles

Topic 16

Confidence Intervals: Proportions

Odd- Numbered Exercise Solutions

Exercise 16-7: Generation M

a. A 95% confidence interval is $.49 \pm (1.96)\sqrt{(.49)(.51)/2032} \approx .49 \pm (1.96)(.0111) \approx .49 \pm .0217 \approx$ (.468, .512).

    You are 95% confident the population proportion of American youths who have a video game player in their bedroom is between .468 and .512.

b. A 95% confidence interval is $.31 \pm (1.96)\sqrt{(.31)(.69)/2032} \approx .31 \pm (1.96)(.01026) \approx .31 \pm .0201 \approx$ (.290, .330).

    You are 95% confident the population proportion of American youths who have a computer in their bedroom is between .290 and .330.

c. The margin-of-error for the 95% CI for the population proportion of American youths who have a video game player in their bedroom is .0217.

    The margin-of-error for the 95% CI for the population proportion of American youths who have a computer in their bedroom is .0201.

d. Yes, a confidence interval's margin-of-error depends on more than its sample size and confidence level. It also depends on the sample proportion. The closer the sample proportion is to .5, the larger the margin-of-error will be. (You see this because both of these confidence intervals had the same confidence level and sample size, but they had different margins-of-error.)

Exercise 16-9: Penny Activities

Results will vary. The following are meant to be representative results:

a. A 95% confidence interval is $.52 \pm (1.96)\sqrt{(.52)(.48)/50} \approx .52 \pm (1.96)(.07065) \approx .52 \pm .1385 \approx$ (.382, .658).

    You are 95% confident the probability a flipped penny will land heads up is between .382 and .658.

b. You calculate $27/50 \pm (1.96)(.07048) \approx .54 \pm .1381 \approx (.402, .678)$.

You are 95% confident the probability a spun penny will land heads up is between .472 and .678.

c. You calculate $32/50 \pm (1.96)(.0679) \approx .64 \pm .1330 = (.507, .773)$.

You are 95% confident the probability a tilted penny will land heads up is between .507 and .773.

d. Based on these results, flipping or spinning a penny could plausibly be 50/50 for landing heads or tails because .50 is contained in both of these confidence intervals, but it is not contained in the confidence interval for a tilted penny.

e. A tilted penny appears to result in the highest probability of landing heads.

Exercise 16-11: Penny Activities

Results will vary. The following are representative results:

a. Setting the margin-of-error to .02 and solving for n:

$$.02 = (1.96)\sqrt{\frac{(.54)(.46)}{n}} \quad \rightarrow \quad n = (.54)(.46)\left(\frac{1.96}{.02}\right)^2 \approx 2485.634$$

This says your sample size should be at least 2,486.

b. Setting the margin-of-error to .02 and solving for n:

$$.02 = (1.96)\sqrt{\frac{(.64)(.36)}{n}} \quad \rightarrow \quad n = (.64)(.36)\left(\frac{1.96}{.02}\right)^2 \approx 2212.762$$

This says your sample size should be at least 2,213.

c. The necessary sample size is largest with the penny flips. In this particular result, the sample proportion of heads is closest to 50% with the penny flips.

Exercise 16-13: Responding to Katrina

a. The margin-of-error is $1.96\sqrt{\frac{(.12)(.88)}{848}} \approx .02187$

A 95% confidence interval is $.12 \pm .0219 \approx (.098, .142)$. You are 95% confident that between 9.8% and 14.2% of all white adults would have answered yes if asked.

b. The margin-of-error is $1.96\sqrt{\frac{(.60)(.40)}{262}} \approx .0593$

A 95% confidence interval is $.6 \pm .0593 \approx (.541, .569)$. You are 95% confident that between 54.1%

and 65.9% of all black adults would have answered yes if asked.

c. The interval for the black adults is more than twice as wide as the interval for the non-Hispanic white adults, but that interval also indicates that the percentage of blacks who believe that race was a factor in the government's slow response is much higher (somewhere between 54–66%) than is the percentage of whites who believe this (only 9.8–14.2%).

d. The black adults have the greater margin-of-error. This is because their group was smaller (the sample size was only about one-fourth the size of the non-Hispanic white adults).

e. You are not told how the sample was selected; in particular, was any attempt made to randomly select these individuals? This is a critical technical condition that you must check in order to determine whether these intervals are valid.

Exercise 16-15: Magazine Advertisements

a. The observational units are the pages of *Sports Illustrated* and *Soap Opera Digest* magazines.

b. *Sports Illustrated*: $54/116 \pm (1.96)\sqrt{(.466)(.534)/116} \approx .466 \pm (1.96)(.0463) \approx .466 \pm .0908 \approx (.375, .556)$.

   Soap Opera Digest: $28/130 \pm (1.96)\sqrt{(.215)(.785)/130} \approx .215 \pm (1.96)(.0361) \approx .215 \pm .0707 \approx (.145, .286)$.

c. You are 95% confident the percentage of all *Sports Illustrated* pages that contain an ad is between 37.5% and 55.6%. Similarly, you are 95% confident the percentage of all *Soap Opera Digest* pages that contain ads is between 14.5% and 28.6%.

d. If you were to repeat this procedure many, many times, always using random samples of 116 pages of *Sports Illustrated* (and 130 pages of *Soap Opera Digest*), 95% of the time you would create intervals that would contain the population proportion of the magazine's pages that contain ads.

e. Yes, each interval contains the sample proportion of pages with ads.

f. This question was silly because the sample proportion will always be the midpoint of the confidence interval.

g. Answers will vary by choice of magazine. Here is a representative answer:

   Newsweek, April 30, 2007: Seventeen of 65 pages contained ads.

   A 95% CI for the population proportion is $17/65 \pm (1.96)\sqrt{(.2615)(.7385)/65} \approx .262 \pm .107 \approx (.155, .369)$

   You are 95% confident that between 15.5% and 36.9% of all *Newsweek* pages contain ads.

Exercise 16-17: Random Babies

Answers will vary. The following is a representative response:

a. The proportion of simulated repetitions that resulted in no mother getting the correct baby was $370/1000 = .370$. (See Exercise 11.1, part k).

b. For a 95% CI, you calculate $.370 \pm (1.96)\sqrt{(.37)(1-.37)/1000} \approx .370 \pm (1.96)(.01527) \approx .370 \pm .0299 \approx (.340, .400)$.

c. You are 95% confident the long-tern proportion of times that no mother would get the correct baby is between .34 and .40.

d. $\pi = .375$.

e. Yes, the 95% confidence interval succeeds in capturing the population parameter.

f. If 1000 different statistics classes carried out this simulation, you would expect roughly 95% or 950 of their intervals to succeed in capturing $\pi = .375$, whereas you would expect about 5% or 50 of these intervals not to contain 375.

g. For an 80% CI, you calculate $.37 \pm (1.282)\sqrt{(.37)(1-.37)/1000} \approx .37 \pm .01957 \approx (.350, .390)$.

Yes, this interval succeeds in capturing $\pi = .375$.

If 1000 different statistics classes carried out this simulation, you would expect roughly 80% or 800 of their intervals to succeed in capturing $\pi = .375$, whereas you would expect about 200 of these intervals not to contain 375.

Exercise 16-19: Marriage Ages

a. The sample size condition is met because $n\hat{p} = 67 > 10$ and $n(1 - \hat{p}) = 27 > 10$. You need to assume this sample was randomly selected, or was at least representative of the population.

b. A 99% confidence interval for the proportion of all marriages in this county for which the *bride* is young than the groom is $.713 \pm (2.567)\sqrt{(.713)(.287)/94} \approx .713 \pm (2.576)(.0467) \approx .713 \pm .120 = (.592, .833)$.

c. A 99% confidence interval for the proportion of all marriages in this county for which the *groom* is young than the groom is $.287 \pm (2.576)(.0467) \approx .287 \pm .120 = (.167, .407)$.

d. Both intervals have the same width (.120), but the centers are different. The interval in part b is centered at .713, whereas the interval in part c is centered at $1 - .713 = .287$.

e. The 99% confidence intervals does suggest that the bride is young than the groom for more than half of the marriages in this county (assuming this sample is representative) because all of the values in the interval are greater than .5

Exercise 16-21: Wrongful Conclusions

a. Andrew's interval (.346, .474) must be incorrect because it is not centered at the sample proportion $\hat{p}$ = .4. (It is centered at .41.)

b. Andrew's sample proportion: $\hat{p}$ = (.682 + .558)/2 = .62

   Becky's sample proportion: $\hat{p}$ = (.611 + .779)/2 = .695

c. Andrew's margin-of-error: (.682 − .558)/2 = .062

   Becky's margin-of-error: (.778 − .611)/2 = .084.

d. In order to solve for the sample size, you would need to know the $z*$ value used in the margin-of-error formula. (You found the margin-of-error in part c by recognizing it as the half-width of the intervals.) Because you do not know the confidence level used, you do not know the critical value ($z*$) used, so you cannot determine which sample size was used by which person.

e. Andrew's critical value: $.062 = z* \sqrt{\dfrac{(.62)(.38)}{100}}$; $\quad \rightarrow \quad \dfrac{.062 \times 10}{\sqrt{(.62)(.38)}} = z* = 1.28.$

   Then using Table II, you find the confidence level used by Andrew is 90%.

   Becky's critical value: $.084 = z* \sqrt{\dfrac{(.695)(.305)}{200}}$; $\quad \rightarrow \quad \dfrac{.084 \times \sqrt{200}}{\sqrt{(.695)(.305)}} = z* = 2.58.$

   Then using Table II, you find the confidence level used by Becky is 99%.

f. Andrew: $\hat{p}$ = (.635 + .533)/2 = .584; $\qquad$ margin-of-error = (.635 − .533)/2 = .051

   Becky: $\hat{p}$ = (.602 + .55)/2 = .576 $\qquad$ margin-of-error = (.602 − .550)/2 = .026

g. Because both researchers are using the same confidence level, you know that the smaller margin-of-error is associated with the larger sample size, so Becky used a sample of size 1000, whereas Andrew using a sample of size 350.

h. Both researchers selected random samples and used a 90% confidence level, so both procedures had a 90% chance of capturing the population proportion.

Exercise 16-23: Penny Thoughts

a. The population parameter is the proportion of all American adults who favor abolishing the penny.

b. For a 95% CI, you calculate

$$.59 \pm (1.96)\sqrt{(.59)(.41)/2316} \approx .59 \pm (1.96)(.01022) \approx .59 \pm .02 = (.57, .61).$$

You are 95% confident that between 57% and 61% of all American adults favor abolishing the penny.

c. The sample size is certainly large enough ($2316 \times .59 \approx 1366 > 10$ and $2416 \times .41 \approx 950 > 10$), but you have no indication of how this sample of adults was selected. If you assume the Harris Poll selected randomly, then the technical conditions are satisfied. Otherwise, you should interpret this interval with caution.

Exercise 16-25: Generation M

a. If the sample size is 2032, you estimate the margin-of-error for a 95% confidence interval to be $1/\sqrt{2032} = .0222$.

b. This approximate error-margin is very close to the margins-of-error found in Exercise 16-7. This approximation is a little closer to the margin-of-error (.0217) for the statistic .49, than for the statistic .31 (margin-of-error = .0201). This is because .49 is closer to .5 than is .31.

Exercise 16-27: Alternative Procedure

Answers will vary. The following is one representative set.

a. The running total is 1620. The percentage of the intervals that succeeded in capturing the actual value of π (.2) is 1620/2000 or 81.0%.

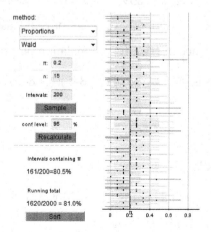

b. If the (usual) Wald confidence interval procedure were working, the percentage found in part a would be close to 95%.

c. Because 81% is not close to 95%, you know the confidence interval procedure is not working well.

d. Using the adjusted Wald procedure, the percentage of intervals that succeeded in capturing the actual value of $\pi = .2$ is 1963/2000 or 98.2%.

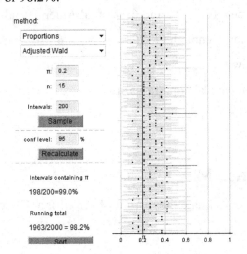

e. Because this value (98.2%) is fairly close to 95% (it's actually larger), this adjusted Wald procedure appears to be working much better than the usual Wald procedure in this case.

Exercise 16-29: Celebrating Mothers

a. The parameter for this study is π = the proportion of all American consumers who planned to celebrate Mother's Day in April 2007.

b. The sample size condition is met because $7859 \times .845 = 6640.9 > 10$ and $7859 \times .164 = 1288.9 > 10$. If you assume this sample was either randomly selected, or is at least a representative sample, then the technical conditions for a confidence interval for a population proportion are met.

c. A 99% confidence interval for π is $.845 \pm (2.576)\sqrt{(.845)(1-.845)/7859} \approx .845 \pm (2.576)(.00408) \approx$ $.845 \pm .01052 \approx (.834, .856)$.

Exercise 16-31: CPR on Pets

a. Let π be the population proportion of all pet owners in 2009. Let $\hat{p}$ be the proportion of all 1166 pet owners who said they would perform CPR on their pet.

A 95% confidence interval for π would be $\hat{p} \pm .029$, so the endpoints of this interval would be $\hat{p} - .209$ and $\hat{p} + .209$.

b. Even if the sample was randomly selected, you do not have enough information to calculate a 95% confidence interval for the population proportion of dog (or cat) owners who say they would perform CPR because you do not know the margin-of-error for just the dog owners, and you cannot calculate this value because you do not know how many of the 1166 pet owners were dog owners.

c. Because only some of the 1166 pet owners surveyed had a dog, the sample size of the dog owners would be less than 1166. This would result in a margin-of-error that is larger than that of all pet owners (.029).

Exercise 16-33: Friday Classes

a. You determine the sample size by setting the margin-of-error to .04 and solving for n:

$$.04 = (1.96)\sqrt{\frac{(.5)(.5)}{n}} \quad \rightarrow \quad n = \left(\frac{(.5)(1.96)}{.04}\right)^2 = 600.25$$

This says your sample size must be at least 601.

b. In order to use 95% confidence, you would have to increase the desired margin-of-error if you wanted to use fewer students in your sample.

c. If you wanted a ±.04 margin-of-error, and you wanted to use fewer students in your sample, you would have to lower the confidence level.

Exercise 16-35: Hanging Toilet Paper

You determine the sample size by setting the margin-of-error to .06 and solving for n:

$$.04 = (1.645)\sqrt{\frac{(.75)(.25)}{n}} \quad \rightarrow \quad n = (.75)(.25)\left(\frac{1.645}{.06}\right)^2 = 111.78; \quad \rightarrow n = 112$$

Topic 17

Tests of Significance: Proportions

Odd- Numbered Exercise Solutions

Exercise 17-7: Properties of *p*-values

a. Yes, it is possible for a *p*-value to be greater than .5. A *p*-value is the probability of obtaining a sample result as or more extreme than (as defined by the alternative hypothesis) the given results by random chance alone if the null hypothesis is true. If the sample result is in the opposite direction from the hypothesized value than conjectured by H_a, then the one-sided *p*-value will be greater than .5. For example, if you are testing H_0: $\pi = .25$, vs. H_a: $\pi > .25$, but you observe $\hat{p} = .2$, the *p*-value = $Pr(\hat{p} > .2$ when $\pi = .25) > .5$. Also, if you have a two-sided alternative and the one-tail probability exceeds .25, then the two-sided *p*-value will be more than .5.

b. No, it is not possible for a *p*-value to be greater than 1 because a *p*-value is a probability and is, therefore, a value between 0 and 1 inclusive.

Exercise 17-9: Penny Activities

You should not be convinced this is not 50–50 process until you know how many times the coin has been flipped in order to obtain this sample value of 75% heads. Suppose the coin has been flipped only four times! This would not be at all convincing. However, obtaining heads 75% of the time in 1000 flips would be very convincing that this is not a 50–50 process.

Using Table II, *p*-value = $Pr(Z > 4.50) < .0002$.

Test decision: Because the *p*-value is very small, reject H_0.

Conclusion in context: You have very strong statistical evidence (*p*-value < .0002) that more than one-fourth of the students at your school would select the right-front tire when asked the "flat-tire" question (as long as this sample is representative of the students at your school on this question).

Exercise 17-11: Feeling Rushed

a. Let π represent the population proportion of all adult Americans who would say that they always feel rushed in 2004.

The null hypothesis is that one-third of all adult Americans would say they always feel rushed. In symbols, the null hypothesis is H_0: $\pi = \frac{1}{3}$.

The alternative hypothesis is that proportion of all adult Americans would say they always feel rushed is not one-third. In symbols, the alternative hypothesis is H_a: $\pi \neq \frac{1}{3}$.

b. The sample was randomly selected and $n\hat{p} = 304 > 10$, $n(1 - \hat{p}) = 673 > 10$, so the technical conditions for conducting this test are satisfied.

c. The sample proportion is $\hat{p} = 304/977 \approx .311$.

Test statistic: $z = \dfrac{.311 - .333}{\sqrt{\dfrac{(.333)(.667)}{977}}} \approx -1.48$

d. Using Table II, p-value $= 2 \times \Pr(Z < -1.48) < 2 \times .0694. = .1388 > .05$

e. The p-value is the probability of finding at least 304 out of a random sample of 977 who say they always feel rushed, assuming the population proportion of all adult Americans who always feel rushed is one-third.

f. Because the p-value is not less than .05, do not reject the null hypothesis at the .05 significance level.

g. You have no convincing evidence (p-value $= .1338$) that the population proportion of adult Americans who report always feeling rushed differs from one-third.

Exercise 17-13: Political Viewpoints

a. In the sample, $497/1309 \approx .38$ consider themselves to be political moderates. This result is clearly greater than 1/3.

b. The null hypothesis is that one-third of all American adults consider themselves to be political moderates. In symbols, the null hypothesis is H_0: $\pi = \frac{1}{3}$.

The alternative hypothesis is that more than one-third of all American adults consider themselves to be political moderates. In symbols, the alternative hypothesis is H_a: $\pi > \frac{1}{3}$.

Technical conditions: The CLT applies here because $1309 \times .333 = 435.1 > 10$, and $1309 \times .667 = 873.1 > 10$, and you have a random sample from the population of interest.

The test statistic is $z = \dfrac{.38 - .333}{\sqrt{\dfrac{(.333)(.667)}{1309}}} \approx 3.58$.

Using Table II, p-value = $\Pr(Z > 3.58) \approx .0000$.

c. Answers will vary by student expectation, but students should expect the p-value to increase because although the observed sample proportion ($124/327 \approx .38$) is the same, the sample size has decreased. This implies the observed sample result is less surprising, corresponding to a larger p-value.

d. To determine the p-value, you calculate

$$z = \frac{.38 - .333}{\sqrt{\dfrac{(.333)(.667)}{327}}} \approx 1.77.$$

Using Table II, p-value = $\Pr(Z > 1.77) = .0384$.

The p-value did indeed increase.

Exercise 17-15: Calling Heads or Tails

Answers will vary. Here is one representative set.

Define the parameter of interest: Let π represent the proportion of all students who would answer "heads" when asked to predict the result of a coin flip.

The null hypothesis is that 70% of all students would answer "heads" when asked to predict the result of a coin flip. In symbols, the null hypothesis is H_0: $\pi = .7$.

The alternative hypothesis is that the proportion of students who would answer "heads" when asked to predict the result of a coin flip is not 70%. In symbols, the alternative hypothesis is H_a: $\pi \neq .7$.

Here is a sketch of the sampling distribution:

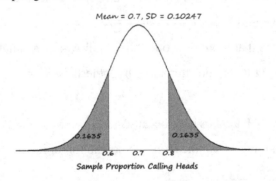

Mean = 0.7, SD = 0.10247

Sample Proportion Calling Heads

Check technical conditions: $20 \times .5 = 20 \times (1 - .5) = 10 \geq 10$. You can consider the sample to be representative of how students would respond to the "predict a coin flip" question.

The test statistic is $z = \dfrac{.8 - .7}{\sqrt{\dfrac{(.7)(.3)}{20}}} \approx 0.98$.

Using Table II, the p-value is $2 \times \mathrm{Pr}(Z > 0.98) = 2 \times .1635 = .325$.

Test decision: Because the p-value is not small, do not reject H_0.

Conclusion in context: Assuming you have a representative sample, you have not found convincing evidence (p-value = .325) that the population proportion of students who will answer "heads" when asked to predict the result of a coin flip differs from .70.

Exercise 17-17: Baseball "Big Bang"

a. The null hypothesis is that the proportion of all major-league baseball games that contain a big bang is .5 In symbols, the null hypothesis is H_0: $\pi = .5$.

The alternative hypothesis is that the proportion of all major-league baseball games that contain a big bang is not .5. In symbols, the alternative hypothesis is H_a: $\pi \neq .5$.

Technical conditions: $968 \times .5 = 968 \times (1 - .5) = 484 > 10$, so this condition is met. You must believe that the games in 1986 are representative of the overall process.

The sample proportion is $419/968 \approx .433$. Here is a sketch of the sampling distribution:

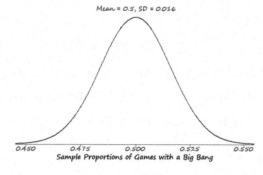

The test statistic is $z = \dfrac{.433 - .5}{\sqrt{\dfrac{(.5)(.5)}{968}}} \approx -4.18$.

Using Table II, p-value $= 2 \times \mathrm{Pr}(Z < -4.18) < 2 \times .0002 = .0004$.

Because the p-value $< .02$, reject the null hypothesis at the $\alpha = .02$ level.

There is very strong statistical evidence (p-value = .0004) that the proportion of all major-league baseball games that contain a big bang is not .5.

b. The null hypothesis is that the proportion of all major-league baseball games that contain a big bang is three-fourths. In symbols, the null hypothesis is $H_0: \pi = .75$.

The alternative hypothesis is that the proportion of all major-league baseball games that contain a big bang is not three-fourths. In symbols, the alternative hypothesis is $H_a: \pi \neq .75$.

Technical conditions: $968 \times .75 = 725 > 10$ and $968 \times (.25) = 242 > 10$, so this condition is met. You must believe that the games in 1986 are representative of the overall process.

The sample proportion is $651/968 \approx .673$. Here is a sketch of the sampling distribution:

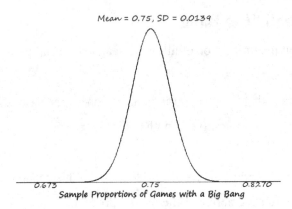

Mean = 0.75, SD = 0.0139

0.673   0.75   0.8270

Sample Proportions of Games with a Big Bang

The test statistic is $z = \dfrac{.673 - .75}{\sqrt{\dfrac{(.75)(.25)}{968}}} \approx -5.56$.

Using Table II, p-value $= 2 \times \Pr(Z < -5.56) < 2 \times .0002 = .0004$.

Because the p-value $< .08$, reject the null hypothesis at the $\alpha = .08$ level.

There is very strong statistical evidence (p-value $= .0004$) that the proportion of all major-league baseball games that contain a big bang is not three-fourths.

Exercise 17-19: Therapeutic Touch

a. The parameter is π, the proportion of times subjects (all therapeutic touch practitioners in such an experimental setup) can correctly identify over which hand the experimenter's hand is held.

The null hypothesis is that the proportion of times subjects can correctly identify over which hand the experimenter's hand is held is .50. In symbols, $H_0: \pi = .5$.

The alternative hypothesis is that the proportion of times subjects can correctly identify over which hand the experimenter's hand is held *is greater than* .50. In symbols, $H_a: \pi > .5$.

b. Note the sample size is large enough; consider these subjects as representative of the performance of all therapeutic touch practitioners.

The sample proportion is $\hat{p} = 123/280 \approx .439$.

The test statistic is $z = \dfrac{.439 - .5}{\sqrt{\dfrac{(.5)(.5)}{280}}} \approx -2.03$.

Using Table II, p-value = $\Pr(Z > -2.03) = 1 - .0212 = .9788$.

c. It makes sense that the p-value is greater .5 in this situation because the sample proportion is less than ½, but you conjectured the population proportion was greater than .5. The subjects were actually less successful that you would expect if they just guessed each time! Such a sample does not provide evidence that subjects can distinguish between the hands *more than half the time*.

d. No, you never "accept" the null hypothesis. You simply have no evidence against the null hypothesis in favor of the alternative, so you continue to assume that .50 is a plausible value of π.

e. You can only conclude that these practitioners showed no evidence of being able to correctly identify more often than not over which hand the experimenter's hand was held.

Exercise 17-21: Hiring Discrimination

a. The parameter of interest (call it π) is the long-run probability of an African American teacher being hired by the City of Hazelwood.

b. The null hypothesis is that the probability of an African American teacher being hired by the City of Hazelwood is .154 (the same as the county proportion). In symbols, the null hypothesis is H_0: $\pi = .154$.

The alternative hypothesis is that the probability of an African American teacher being hired by the City of Hazelwood is less than .154. In symbols, the alternative hypothesis is H_a: $\pi < .154$.

Here is a sketch of the sampling distribution:

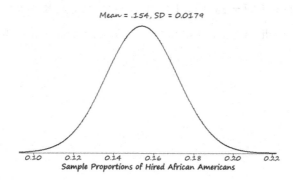

Mean = .154, SD = 0.0179

Sample Proportions of Hired African Americans

Technical conditions: $405 \times .154 = 62.37 > 10$, and $405 \times .846 = 342.63 > 10$, so the sample size condition is met. This is hardly a random sample, but you can proceed with caution if you consider this sample to be representative of the overall hiring process.

The sample proportion is $\hat{p} = 15/405 \approx .037$.

The test statistic is $z = \dfrac{.037 - .154}{\sqrt{\dfrac{(.154)(.846)}{405}}} \approx -6.52$.

Using Table II, p-value $= \Pr(Z < -6.52) < .0002$.

With such a small p-value, reject H_0 at any commonly used significance level.

Based on these data, you have overwhelming statistical evidence (p-value $< .0002$) that the probability of an African American being hired by the school district is less than 15.4%.

c. The null hypothesis is that the probability of an African American teacher being hired by the City of Hazelwood is .057. In symbols, the null hypothesis is H_0: $\pi = .057$.

The alternative hypothesis is that the probability of an African American teacher being hired by the City of Hazelwood is less than .057. In symbols, the alternative hypothesis is H_a: $\pi < .057$.

Here is a sketch of the sampling distribution:

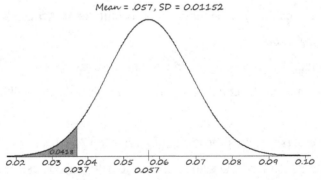

Mean = .057, SD = 0.01152

Sample Proportions of Hired African Americans

Technical conditions: $405 \times .057 = 23.09 > 10$, and $405 \times .943 = 381.92 > 10$, so the sample size condition is met. This is hardly a random sample, but you can proceed with caution if you consider this sample to be representative of the overall hiring process.

The sample proportion is $\hat{p} = 15/405 = .037$.

The test statistic is $z = \dfrac{.037 - .057}{\sqrt{\dfrac{(.057)(.943)}{405}}} \approx -1.73$.

Using Table II, p-value = $\Pr(Z < -1.73) < .0418$.

Because the p-value = $.0418 > .01$, do not reject H_0 at the $\alpha = .01$ significance level.

You do not have sufficiently strong evidence (at the .01 level) to conclude that the proportion of African Americans hired by the city of Hazelwood School District is less than .057.

d. There is very strong statistical evidence that the hiring rate of African Americans is below that of the entire county, but this hiring rate may be similar to the population proportion of African Americans in the county if the city of St. Louis is excluded.

Exercise 17-23: Veterans' Marital Problems

Define the parameter of interest: Let π represent the proportion of all Vietnam veterans who are divorced.

The null hypothesis is that the proportion of all Vietnam veterans who are divorced is .27. In symbols, the null hypothesis is H_0: $\pi = .27$.

The alternative hypothesis is that the proportion of all Vietnam veterans who are divorced is more than .27. In symbols, the alternative hypothesis is H_a: $\pi > .27$.

Check technical conditions: $2101 \times .27 = 576.27$, and $2101 \times .73 = 1533.73$ which both exceed 10. You are not told how the sample of veterans was selected.

The sample proportion is $\hat{p} = 777/2101 \approx .37$.

Test statistic: $z = \dfrac{.37 - .27}{\sqrt{\dfrac{(.27)(.73)}{2101}}} = 10.31$.

Using Table II, p-value = $\Pr(Z > 10.31) < .0002$.

Test decision: With such a small p-value, reject H_0 at any commonly used significance level.

Conclusion in context: You have overwhelming statistical evidence (p-value $< .0002$) that the proportion of all Vietnam veterans who are divorced is more than 27%. If the divorce rate for Vietnam veterans were really 27%, you could expect to see a sample result like this (777/2101 or more divorced veterans) in less than .02% of random samples. Because this sample result would almost never happen if the null hypothesis were true, you have strong evidence that your null hypothesis is false and 27% is not a plausible value for the divorce rate of Vietnam veterans. (You conclude that this rate is greater than 27%.)

Exercise 17-25: Monkeying Around

Define the parameter of interest: Let π represent the proportion of rhesus monkeys who would approach the targeted box.

The null hypothesis is that half of rhesus monkeys would approach the targeted box. In symbols, the null hypothesis is H_0: $\pi = .5$.

The alternative hypothesis is that more than half of all rhesus monkeys would approach the targeted box. In symbols, the alternative hypothesis is H_a: $\pi > .5$.

Check technical conditions: $40 \times .5 = 40 \times (1 - .5) = 20 > 10$. You need to assume the rhesus monkeys were randomly selected.

Test statistic: $z = \dfrac{.75 - .5}{\sqrt{\dfrac{(.5)(.5)}{40}}} = 3.16$.

Using Table II, p-value $= \Pr(Z > 3.16) = .0008$.

Test decision: With such a small p-value, reject H_0 at the $\alpha = .05$ significance level.

Conclusion in context: You have very strong statistical evidence (p-value $= .0008$) that more than half of all rhesus monkeys would approach the targeted box.

Exercise 17-27: Stating Hypotheses

a. The null hypothesis is that the population proportion of defective items in this production process is .005. In symbols, the null hypothesis is H_0: $\pi = .005$.

   The alternative hypothesis is that the population proportion of defective items in this production process is less than .005. In symbols, the alternative hypothesis is H_a: $\pi < .005$.

b. The null hypothesis is that the probability of this amateur bowler bowling a strike is .20. In symbols, the null hypothesis is H_0: $\pi = .20$.

   The alternative hypothesis is that the probability of this amateur bowler bowling a strike is greater than .20. In symbols, the alternative hypothesis is H_a: $\pi > .20$.

c. The null hypothesis is that the population proportion of students at this college who have at least one class on Friday is .60. In symbols, the null hypothesis is H_0: $\pi = .60$.

The alternative hypothesis is that the population proportion of students at this college who have at least one class on Friday is not .60. In symbols, the alternative hypothesis is H_a: $\pi \neq .60$.

Exercise 17-29: Matching Pets to Owners

a. Let π be the probability of correctly matching the professor with her cat.

The null hypothesis is that the probability of correctly matching this professor with her cat is one-third. In symbols, the null hypothesis is H_0: $\pi = \frac{1}{3}$.

The alternative hypothesis is that the probability of correctly matching this professor with her cat is greater than one-third. In symbols, the alternative hypothesis is H_a: $\pi > \frac{1}{3}$.

b. Sample size technical condition: $34 \times \frac{1}{3} = 11.32 > 10$ and $34 \times \frac{2}{3} = 22.68 > 10$. This condition is satisfied.

c. The sample proportion is $15/34 \approx .441$.

The test statistic is $z = \dfrac{.441 - .333}{\sqrt{\dfrac{(.333)(667)}{34}}} \approx 1.34$.

Using Table II, the p-value is $\Pr(Z > 1.34) = .0901$.

d. The p-value is the probability of obtaining at least 15 out of a random sample of 34 students who correctly match the professor with her cat, assuming that the probability of a correct match is one-third (the students are just guessing about which cat is the professor's).

e. Because the p-value = 0901 < .10, you reject the null hypothesis at the $\alpha = .10$ significance level. However, the p-value > .01, so you would not reject the null hypothesis at the $\alpha = .01$ significance level.

f. You have found weak statistical evidence ($p = .0901$) that these students are able to correctly match this professor with her cat better than they could by just guessing. This evidence is not overly strong.

Topic 18

More Inference Considerations

Odd- Numbered Exercise Solutions

Exercise 18-7: Charitable Contributions

a. For a 90% CI, you calculate $.788 \pm (1.645)\sqrt{.788(1-.788)/1334} \approx (.769, .806)$.

b. For a 99% CI, you calculate $.788 \pm (2.576)\sqrt{.788(1-.788)/1334} \approx (.759, .817)$.

c. Yes, the sample proportion does differ from .75 at the $\alpha = .01$ significance level because .75 is *not* in the 99% confidence interval. Thus, you *would reject* the null hypothesis H_0: $\pi = .75$ vs. the two-sided alternative H_a: $\pi \neq .75$, and would decide that .75 is not a plausible value for π.

d. The sample proportion does *not* differ from .80 at the $\alpha = .10$ significance level because .80 *is* in the 90% confidence interval. Thus, you *would not reject* the null hypothesis H_0: $\pi = .80$ in favor of the two-sided alternative H_a: $\pi \neq .80$. With 90% confidence, .80 seems to be a plausible value for π.

Exercise 18-9: Distinguishing Between Colas

a. The null hypothesis is that the subject is just guessing and would correctly identify the brand of cola one-third of the time in the long run. In symbols the null hypothesis is H_0: $\pi = \frac{1}{3}$.
   The alternative hypothesis is that the subject would correctly identify the different brand of cola more than one-third of the time in the long run. In symbols, the alternative hypothesis is H_a: $\pi > \frac{1}{3}$.

b. You can't tell whether Randy's sample data would necessarily lead to rejecting the null hypothesis because you don't know how many times he ran the experiment. If he only tried twice and succeeded 50% of the time, this would not be convincing evidence that he is doing better than guessing, but if he tried 500 times and succeeded 50% of the time, this would be very convincing evidence.

c. If $n = 50$, and $\hat{p} = .46$, the test statistic is $z = \dfrac{.46 - .3333}{\sqrt{\dfrac{(.3333)(.6667)}{50}}} \approx 1.90$ and the p-value is $\Pr(Z < 1.90)$

   $= .0287 < .05$.

Yes, at the .05 significance level, reject H_0 and conclude that Randy is doing better than guessing.

d. Assuming Randy's success rate is 50%, the CLT says $\Pr(\hat{p} > .46) = .7142$ with samples of size 50.

 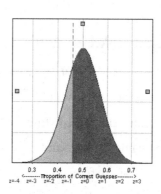

So, you expect about 71% of random samples to yield a sample proportion greater than .46 if Randy's success rate is 50%.

e. The probability will increase because with the larger sample size the standard deviation will decrease from .0707 to .05.

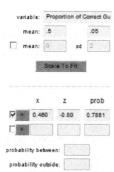 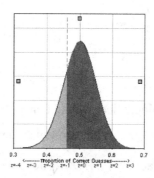

With the larger sample size, you expect more sample proportions to be close to .5 and therefore greater than .46.

f. The probability will increase because the standard deviation will decrease to .06665 and (more importantly) because .46 will be so much further below ⅔ than it is below ½.

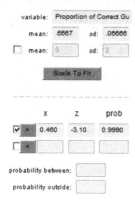 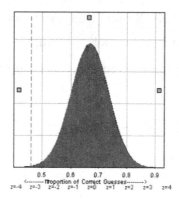

If the population proportion is even further above .46, the probability of obtaining a sample proportion of .46 or more quickly increases.

Exercise 18-11: Phone Book Gender

a. Ignoring the initials, $77 + 14 + 2 \times 36$ or 163 first names are listed in the phone book; $(14 + 36)/163$ or $50/163 \approx .307$ of those names are female.

b. The observational units in this study are listings in the San Luis Obispo County telephone book.

c. The population is all first names in the San Luis Obispo County phone book, and the parameter is the proportion of those names that are female.

d. The null hypothesis is that half the first names in the San Luis Obispo County phone book will be female. In symbols, the null hypothesis is $H_0: \pi = .5$.

   The alternative hypothesis is that less than half of the first names in the San Luis Obispo County phone book will be female. In symbols, the alternative hypothesis is $H_a: \pi < .5$.

   The test statistic is $z = \dfrac{.307 - .5}{\sqrt{\dfrac{(.5)(.5)}{163}}} \approx -4.93$.

   Using Table II, p-value $= \Pr(Z < -4.93) \approx .0000$.

   With this very small p-value, reject H_0 and conclude that you have very strong statistical evidence that less than half of the first names in the San Luis Obispo County phone book are female.

e. For a 95% CI, you calculate $.307 \pm (1.96)\sqrt{.307(1 - .307)/163} \approx (.236, .377)$.

   You are 95% confident the proportion of female first names in the San Luis Obispo County phone book is between .236 and .377.

f. Because the data used to create this confidence interval and run this significance test were not collected from a simple random sample and are not likely to be representative of the proportion of women living in San Luis Obispo County, you should not use this data to try to determine whether half of the residents of San Luis Obispo County are female. Note that living in the county is very different from being listed as a female in the county phone book.

Exercise 18-13: Racquet Spinning

a. $H_0: \pi = .5$ $H_a: \pi < .5$

b. The test statistic is $z = \dfrac{.46 - .5}{\sqrt{\dfrac{(.5)(.5)}{100}}} = -0.80$.

Using Table II, p-value = $Pr(Z < -0.80)$ = .2119.

This is the same test statistic that you found for the two-sided test, but the p-value is half as large (see Exercises 18-12 and 17-28)

c. The test statistic is $z = \dfrac{.54 - .5}{\sqrt{\dfrac{(.5)(.5)}{100}}} = 0.80$.

Using Table II, p-value = $Pr(Z < 0.80)$ = .7881.

The test statistic is the opposite of the test statistic value found in the one-sided and two-sided tests when $\hat{p}$ was 46/100. The p-value is much greater than either of the p-values found when $\hat{p}$ was .46.

d. The formal test of H_a: $\pi < .5$ is not necessary when the sample proportion ($\hat{p} = .54$) is greater than .5. You know this sample proportion will not be able to provide evidence that $\pi < .5$. Because the observed result is in the wrong direction, the p-value will exceed .5.

Exercise 18-15: Penny Activities

a. A 95% CI for flips is $14{,}709/29{,}015 \pm (1.960(.002931)) \approx (.5012, .5127)$.

A 95% CI for spins is $9{,}197/20{,}422 \pm (1.960(.003482)) \approx (.4435, .4572)$.

A 95% CI for tilts is $10{,}087/14{,}611 \pm (1.960)(.003825) \approx (.683, .694)$.

b. These intervals are so narrow because the sample sizes are so large (which makes the margins-of-error very small).

c. You would reject H_0: $\pi = .5$ vs. H_a: $\pi \neq .5$ at the .05 significance level for all three situations (flips, spins, and tilts) because .5 is not contained in any of the 95% confidence intervals.

d. H_0: $\pi = .5$ vs. H_a: $\pi \neq .5$

For flips, the test statistic is $z = \dfrac{.5069 - .5}{\sqrt{\dfrac{(.5)(.5)}{29015}}} \approx 2.37$.

Using Table II, p-value = $2 \times Pr(Z > 2.37) = 2 \times .0089 = .0178 < .05$

With the p-value less than .05, reject H_0; the statistical evidence suggests that the probability of obtaining a head with a *flipped* penny is not .5 (at the .05 significance level).

For spins, the test statistic is $z = \dfrac{.450 - .5}{\sqrt{\dfrac{(.5)(.5)}{20422}}} \approx -14.19$.

p-value $= 2 \times \Pr(Z < -14.19) \approx 2 \times .0000 \approx .0000 < .05$

Reject H_0; there is very strong statistical evidence that the probability of obtaining a head with a *spun* penny is not .5 (at the .05 significance level).

For tilts, the test statistic is $z = \dfrac{.690 - .5}{\sqrt{\dfrac{(.5)(.5)}{14611}}} \approx 46.02$.

Using Table II, p-value $= 2 \times \Pr(Z > 46.02) \approx 2 \times .0000 \approx .0000 < .05$

Reject H_0; with the very large test statistic, there is overwhelming statistical evidence that the probability of obtaining a head with a *tilted* penny is not .5 (at the .05 significance level).

Yes, these results agree with your answer to part c.

e. No; although you are confident that the probability of obtaining a head with a flipped penny is not .50, the probability is quite close to .50. The confidence interval tells you this probability is somewhere between .5012 and .5127, which for all *practical* purposes is .50.

Exercise 18-17: Hypothetical Baseball Improvements

Answers will vary. The following is based on one representative use of the applet:

From the simulation, the approximate power is 84/200 = .42 (using 30 at-bats, alternative value of π = ⅓).

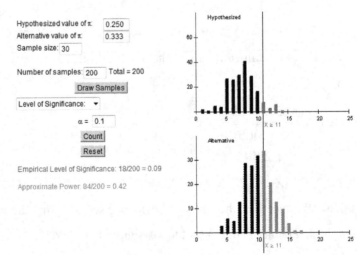

If you use a greater significance level, the power of the test will increase. When the significance level is .05, the player needs 13 hits in 30 at-bats in order to demonstrate his improvement, and the power of the test is roughly 14%, which means it is unlikey his improvement will be noticed. However, if you use a increased significance level of .10, the player needs fewer hits (11 in this case) in order to demonstrate his improvement, and the power of the test becomes approximately 42%; his chances of displaying his improvement in 30 at-bats have increased signficantly.

Exercise 18-19: Emotional Support

a. Hite margin-of-error: $(1.96)\sqrt{\dfrac{(.96)(.04)}{4500}} = .005726$

ABC News/*Washington Post* margin-of-error: $(1.96)\sqrt{\dfrac{(.44)(.56)}{767}} \approx .03514$

Hite 95% CI: $.96 \pm .005726 \approx (.954, .966)$

ABC News/*Washington Post* 95% CI: $.44 \pm .03513 \approx (.405, .475)$

b. No, these two intervals are not similar. They do not overlap at all and their widths are quite different.

c. The Hite survey has the smaller margin-of-error (and the narrower confidence interval).

d. You have much more confidence (95%) in the ABC News/*Washington Post* confidence interval because their sample proportion was obtained from a random sample, unlike Hite's. Because Hite did not use a random sample, you really cannot make any confidence level statements for her interval.

Exercise 18-21: Hiring Discrimination

A Type I error would be deciding there is discrimination (concluding that not enough African Americans are hired by the Hazelwood School District) when, in fact, there is no discrimination. A Type II error would be failing to realize that the Hazelwood School District is discriminating in their hiring practices (not hiring enough African Americans). In this situation, which type of error is more serious is a personal opinion (false accusations vs. allowing discrimination to continue).

Exercise 18-23: Friend or Foe

a. The proportion of infants in this sample who chose the helper toy is 14/16 or .875. This is larger than one-half, as the researchers expected.

b. The parameter of interest (call it π) is the proportion of all 10-month-old infants who would choose the helper toy.

c. The null hypothesis is that the proportion of infants who would choose the helper toy is .5 (infants have no preference for either type of toy). In symbols the null hypothesis is H_0: $\pi = .5$.
    The alternative hypothesis is that the proportion of infants who would choose the helper toy is more than .5 (infants have a preference for the helper toy). In symbols, the alternative hypothesis is H_a: $\pi >$ .5.

d. Because $n\pi = 16(.5) = 8 < 10$, the technical conditions for the one-proportion z-test are not satisfied.

Exercise 18-25: Friend or Foe

a. The probability is .002 that you would find a sample result of 14 or more heads in 16 tosses assuming that heads and tails are equally likely outcomes. This is equivalent to saying the probability is .002 that you would find 14 or more of sixteen infants choosing the "helper" toy, assuming infants have no real preference for either toy.

b. Based on this p-value, you have strong statistical evidence that infants do have a preference for the "helper" toy. Because a result as, or more extreme, as the one found by the researchers has only a .2% probability of occurring by random chance alone if the infants have no preference for either toy, the result found by the researchers convinces you that infants prefer the helper toy.

Exercise 18-27: Racquet Spinning

a. A Type I error would be deciding that "up" and "down" are not equally likely positions when, in fact, they are.

b. A Type II error would be failing to realize that the positions "up" and "down" are not equally likely to occur.

c. Because you failed to reject the null hypothesis in part a of Exercise 18-26, if you made an error, it could only be a Type II error. A Type II error occurs when you fail to reject the null hypothesis (but should reject). You could not make a Type I error in this case because you did not reject the null hypothesis.

Exercise 18-29: Seat Belt Use

a. The population of interest is all drivers who entered convenience stores in El Paso, Texas in 2001.

The parameter is the proportion of all these drivers who would report that they always wear seat belts.

b. Let π = the proportion of all the drivers in the population who would report always wearing a seat belts.

A 99% confidence interval for π is $.75 \pm 2.576\sqrt{(.75)(.25)/612} = (.705, .795)$.

c. No, .615 is not in the 99% CI.

d. The sample was not selected randomly and may not be representative of the population. Therefore the confidence interval procedures are not likely to be valid in this situation. In addition, estimating the proportion of drivers who *report* always wearing a seat belt is very different from estimating the proportion of drivers who do, in fact, always wearing a seat belt. Drivers may not be willing to report that they occasionally or rarely wear a seat belt, or they may think they wear a seat belt more often than they do.

e. A 99% confidence interval for the proportion of drivers in this population who actually wear a seat belt is $.615 \pm 2.576\sqrt{(.615)(1-.615)/612} = (.564, .666)$.

Topic 19

Confidence Intervals: Means

Odd- Numbered Exercise Solutions

Exercise 19-7: Body Temperatures

a. Here are a histogram and normal probability plot of the data:

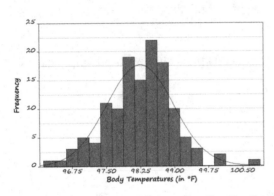 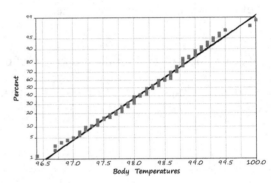

These data appear to be fairly normally distributed.

b. The normality of the *population* of body temperatures is not required for this *t*-procedure to be valid in this case because the sample size is large (130 > 30).

c. You are not told how the sample was selected, so in order for this *t*-procedure to be valid, you will need to assume these 130 adults were randomly selected from the population, or that they are representative of the population of health adults.

d. Using Table III with 100 degrees of freedom, a 95% confidence interval for the population mean body temperature is $98.249 \pm (1.984)(0.733) / \sqrt{130} = 98.249 \pm (1.984)(.06429) = 98.249 \pm .1275 = (98.12, 98.38)$°F.

Using the applet, a 95% CI is (98.073, 98.327) °F.

e. You are 95% confident the population mean body temperature is between 98.12°F and 98.38°F. By "95% confident" you mean that if this procedure was used on thousands of random samples from the same population, roughly 95% of those intervals would succeed in capturing the population mean body temperature.

f. No, it does not appear that 98.6°F is a plausible value for the mean body temperature of all healthy adults because this value does not appear in the confidence interval.

g. Answers will vary by student expectation, but the width of the confidence interval will increase because the sample size has decreased.

h. Using Table III with 100 degrees of freedom, a 95% confidence interval for the population mean body temperature is $98.249 \pm (1.984)(0.733) / \sqrt{130} \approx 98.249 \pm (1.984)(.2033) \approx 98.249 \pm .40334 \approx (97.12, 98.38)°F$.

Using the applet, a 95% CI is $(97.503, 98.995)°F$.

This confidence interval has the same midpoint as the interval created in part d ($\bar{x} = 98.249°F$), but it is much wider (.81°F vs. .36°F).

98.6°F. You are 95% confident the average body temperature of healthy adult females is between

Exercise 19-9: Social Acquaintances

Answers will vary. The following are one representative set of answers:

a. $\bar{x} = 48.03$ people, $s = 25.25$ people, $n = 35$, df = 34.

For a 90% CI, you calculate $40.83 \pm (1.691)(25.25 / \sqrt{35}) = (33.6128, 48.0472)$ people

b. You are 90% confident the population mean number of acquaintances of all students at your school is between 33.6 and 48 people. In this context, 90% confident means that if you were to create similar intervals over and over again, using random samples of 35 students from your school, in the long run approximately 90% of the intervals you create would contain the unknown population mean number of acquaintances of all students at your school, and 10% of the intervals would not contain this value μ. You will never know for certain whether a particular interval does, or does not, contain μ.

c. For this class, 10/35 = .28571 of the sampled students have an acquaintance number that falls within this confidence interval. (See Exercise 9-8.)

d. This percentage (28.6%) is not at all close to 90%, and there is no reason that it should be because the confidence interval only tells you roughly what the population *mean* should be. It does not tell you what the individual sample values should be.

Exercise 19-11: Nicotine Lozenge

a. The population is all smokers who wish to quit smoking. The parameter of interest is μ, mean number of cigarettes smoked per day by the population of smokers.

b. For a 99% CI, you calculate $22 \pm (2.576)(10.8)/\sqrt{1818} = (21.3475, 22.6525)$ cigarettes per day with degrees of freedom equal to infinity, or $22 \pm (2.586)(10.8)/\sqrt{1818} = (21.34, 22.66)$ cigarettes per day with 500 degrees of freedom.

c. No; based on this interval, it does not seem plausible to assert that the population mean is 20 cigarettes per day because 20 is not contained in this confidence interval. It appears that the average number of cigarettes smoke per day is slightly more – at least 21.3 cigarettes per day.

Exercise 19-13: Critical Values

a. Here is the completed table:

Degrees of Freedom	Confidence Levels			
	80%	90%	95%	99%
4	1,533	2.132	2.776	4.604
11	1.363	1.796	2.201	3.106
223	1.319	1.171	2.069	2.807
80	1.292	1.664	1.990	2.639
Infinity	1.282	1.645	1.960	2.576

b. The critical value t^* gets larger as the confidence level increases.

c. The critical value t^* gets smaller as degrees of freedom increase.

d. Yes, the critical values from the t-distribution corresponding to infinitely many degrees of freedom are the z^* critical values.

Exercise 19-15: Coin Ages

Answers will vary. The following is one representative set of answers.

a. Using line 7 of the Random Digits Table, you select coins numbered 835, 944, 872, 096, 632, 397, 245, 031, 891, and 370, which have ages 22, 29, 23, 1, 15, 7 4 0, 19, and 6 years, respectively.

b. For a 90% CI, with 9 degrees of freedom, you calculate $12.6 \pm (1.833)(10.3/\sqrt{10}) = (6.63, 18.57)$ years.

c. Technical conditions: No, the technical conditions for this procedure have not been met. The sample was selected randomly, but it was small $(n < 30)$, and the population definitely is not normally distributed. It has a very strong skew to the right. Therefore a confidence interval statement could be inaccurate.

d. The interval does succeed in capturing the population mean $\mu = 12.26$ years. (This may not always be the case, however.)

e. Yes, if you planned to construct a 95% confidence interval, this method would be more likely to capture the population mean than the 90% confidence interval method because the resulting interval would be wider.

f. If you planned to take a random sample of 40 pennies instead of 10 pennies, the 90% confidence interval procedure method would be more appropriate (the technical conditions for the procedure to be valid would be met as the sample size would now be large enough). However, assuming both methods are valid, it would still be a 90% confidence interval so, in the long run, this interval method has the same chance of capturing the parameter as the method used in part b. But the interval would probably be less wide (the sample mean of 40 pennies is more likely to be close to the population mean than the sample mean of 10 pennies.

Exercise 19-17: Close Friends

a. The observational units are the adult Americans who were interviewed by the GSS. The variable is the *number of close friends that an adult American has*. This variable is quantitative.

b. A *t*-interval is valid in spite of the strong right skew because the sample size is very large (1467).

c. For a 90% confidence interval you calculate $1.987 \pm (1.645)(1.7708 / \sqrt{1467}) = (1.911, 2.0631)$ friends, with degrees of freedom equal to infinity.

d. The reasonable interpretations of this interval are:
You can be 90% confident that the mean number of close friends in the population is between the endpoints of this interval.
If you repeatedly took random samples of 1467 people and constructed t-*intervals in this same manner, 90% of the intervals in the long run would include the population mean number of close friends.*

e. *Ninety percent of all people in this sample reported a number of close friends within this interval* is incorrect because a confidence interval claims to capture the population mean, not individual members of the sample.

If you took another sample of 1467 people, there is a 90% chance that its sample mean would fall within this interval is incorrect because you are not trying to capture sample means; you are trying to capture the population mean. This statement would be correct if the interval had been constructed around the population mean rather than the sample mean.

If you repeatedly took random samples of 1467 people, this interval would contain 90% of your sample means in the long run is not a correct interpretation because you cannot predict how many of the other sample means it would contain—the interval procedure is estimating the population mean. You are not saying other sample means should be within two standard deviations of the one you observed, but that sample means in general should fall within two standard deviations of the actual population mean.

It is incorrect to say, "*This interval captures the number of close friends for 90% of the people in the population*" because this interval estimates the mean number of friends—not the number of individual friends for any person.

f.   If the sample size were larger, the interval would have the same midpoint, but would be narrower. If the sample mean were larger, the interval would have a larger midpoint, but would have the same width.

If the sample values were less spread out, the standard deviation would be smaller, so the margin-of-error would be smaller, so the interval would be narrower (but would have the same midpoint).

If every person in the sample reported one more close friend, the sample mean would be greater (by 1), so the midpoint of the interval would increase by 1, but the width would be unchanged.

Exercise 19-19: Sleeping Times

a.   This is a legitimate interpretation of the interval.

b.   This is a legitimate interpretation of the interval.

c.   This is not a legitimate interpretation of the interval. The interpretation is not technically correct because μ is not random. It has some fixed (but unknown) value, so μ is either inside the interval you produced or it is not; it is not sometimes between the two numbers you calculated and sometimes not between those two numbers. (See the Watch Out comments at the end of Topic 16.)

d.   This is not a legitimate interpretation of the interval; the confidence interval is attempting to estimate the average student sleep time. It says nothing about any individual student's sleep time.

e.   This is not a legitimate interpretation of the interval; the confidence interval is attempting to estimate the average student sleep time in the entire population. It says nothing about any individual student's sleep time, so it cannot predict what percentage of the students' sleep times (in the population) will fall in this interval.

Exercise 19-21: Hypothetical ATM Withdrawals

a. Although all three distributions have the same number of withdrawals (50), the same midpoint ($70), and the same standard deviation ($30.30), their dotplots are very dissimilar. There were only two distinct amounts withdrawn from the first ATM, $40 and $100, and these amounts were withdrawn in equal proportions. There were exactly three amounts withdrawn from the third ATM: $20 and $120 were withdrawn about nine times each, whereas $70 was withdrawn about three times as often from this ATM. There were many different amounts withdrawn from the second ATM. Every other amount ($20, $40, $60, $80, $100, and $120) was withdrawn infrequently (only twice), whereas the remaining amounts ($30, $50, $70, $90, and $110) tended to be withdrawn about four times more often.

b. Here is the completed table:

	Sample Size	Sample Mean	Sample SD	95% CI for μ
Machine 1	50	$70	$30.30	($61.3875, $78.6125)
Machine 2	50	$70	$30.30	($61.3875, $78.6125)
Machine 3	50	$70	$30.30	($61.3875, $78.6125)

c. This activity clearly reveals that a confidence interval for a mean does not display all aspects of a distribution. In this case, you had three distributions that looked very different, but had the same sample sizes, centers, and standard deviations, and thus the same confidence intervals.

Exercise 19-23: Credit Card Usage

a. You need to know the sample standard deviation.

b. For a 95% CI, you calculate $\$2169 \pm (1.965)(\$1000 / \sqrt{1074}) = (\$2109.04, \$2228.96)$ using Table III with 500 degrees of freedom. With technology: ($2109.10, $2228.90).

   You are 95% confident the average credit card balance in the population of all undergraduate students who held a credit card in 2004 was between $2109 and $2229.

c. No, you should not expect 95% of the population to have a credit card balance in this interval. This interval predicts the *average* balance of credit card holders; it says nothing about the balance of individual credit card holders.

Exercise 19-25: Television Viewing Habits

a. No, the fact that the sample data are strongly skewed to the right does not indicate that the technical conditions for a *t*-test are not satisfied. The *t*-procedures are valid because the sample size is large $(1324 > 30)$, and the sample was randomly selected.

b. You calculate $2.982 \pm (1.96)(2.659 / \sqrt{1324}) = 2.982 \pm .1432 = (2.839, 3.125)$ with degrees of freedom equal to infinity or $2.982 \pm (1.965)(2.659 / \sqrt{1324}) = 2.982 \pm .1436 = (2.84, 3.13)$ with 500 degrees of freedom.
You are 95% confident the average number of hours of television watched by adult Americans is between 2.84 and 3.13 hours per day.

c. The proportion of sample data that fall within this interval is (215)/1324 or .162 (See Exercise 8-28). This percentage (16.2%) is *not* close to 95%, but there is no reason that it should be. The confidence interval estimates the *average* number of hours of television watched per day; it does not tell you anything about how much television any individual watches per day.

Exercise 19-27: Hockey Goals

a. The standard error of the sample mean (.377) is also equal to $s / \sqrt{n} = 2.472 / \sqrt{43}$ goals per game.

b. You are 95% confident the average number of goals scored during the NHL 2009-2010 season was between 4.797 and 6.319 goals per game.

c. If you had calculated this confidence interval by hand, you could have used $t^* = 2.021$ with 40 degrees of freedom from Table III, or $t^* = 2.01954$ with 42 degrees of freedom (using Minitab).

d. The right skew of the distribution of goals in this sample does not invalidate this confidence interval. If the sample is clearly non-normal, you need the sample size to be large $(n \geq 30)$, and it is in this case $(n = 43)$.

e. No, the fact the number of goals scored in only 10 of these games falls within the 95% confidence interval does not call the validity of the interval into question. The interval predicts the average number of goals scored per game for the season; it does not predict the number of goals scored in any particular game, so you should have no expectations about how many of these games will have values that fall in the CI.

Exercise 19-29: Birth Weights

Answers will vary by student choice of variable. The table below provides summary statistics for each of these variables.

Variable	n	Mean	SD	Min.	Q_L	Median	Q_U	Max.
Father'sAge	420	30.021	6.651	16	25	30	34	50
Mother's Age	500	26.882	6.354	13	22	26	32	50
Weeks of Gestation	499	38.333	3.012	20	38	39	40	45
Number of Pre-natal Visits	498	12.207	3.922	0	10	12	15	30
Mother's Weight Gain	497	30.4	14.177	0	20	30	40	75

Father's age:

a. The ages of the fathers are slightly skewed to the right with a mean of 30.02 years and a standard deviation of 6.651 years. Only twelve of these 420 fathers were less than 20-years-old at the time of these births, and most of the fathers were between 25 and 37-years-old. The oldest father was 50-years-old, and there were fathers of every age from 16- to 48-years-old.

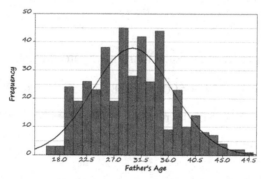

b. Yes, the technical conditions for a *t*-confidence interval are satisfied by these data. The sample was randomly selected from the population of all babies born in North Carolina in 2005, and although these data are not normally distributed, the sample size is large ($n = 420 > 30$). Note that this answer will be the same for any of these variables, regardless of the shape of the data, because each of these variables had a sample size of at least 420.

c. Using Table III with degrees of freedom equal to infinity, you calculate

$$30.021 \pm (1.96)(6.651 / \sqrt{420}) = 30.021 \pm .63609 = (29.38, 30.66) \text{ years.}$$

Using Table III with 100 degrees of freedom, you calculate

$$30.021 \pm (1.984)(6.651/\sqrt{420}) = 30.021 \pm .64388 = (29.38, 30.66) \text{ pounds.}$$

You are 95% certain the average age of the fathers of all babies born in North Carolina in 2005 was between 29.38 years and 30.66 years.

Mother's age:

a. The ages of the mothers are skewed to the right with a mean of 26.88 years and a standard deviation of 6.354 years (slightly less than the standard deviation of the father's ages). The youngest mother was only 13, and there were a 14- and two 15-year-old mothers. The oldest mother was 50, and there were eleven mothers who were in their early forties when they gave birth in 2005. The vast majority of the mothers were aged 20 to 34, with only 12.8% of the mothers being teen-agers.

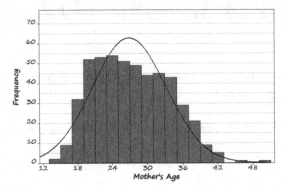

b. See above.

c. Using Table III with degrees of freedom equal to infinity, you calculate

$$26.882 \pm (1.96)(6.354/\sqrt{500}) = 26.882 \pm .5570 = (26.33, 27.44) \text{ years.}$$

You are 95% certain the average age of the mothers of all babies born in North Carolina in 2005 was between 26.33 years and 27.44 years.

Completed weeks of gestation:

a. The histogram of the number of weeks to gestation is strongly skewed to the left with low outliers at 20, 22, 24, 25, and 26 weeks. The vast majority of these gestation times were between 37 and 40 weeks, with more than a quarter of them lasting 39 weeks. The longest pregnancy in this sample lasted 45 weeks.

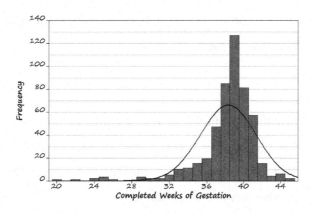

b. See above.

c. Using Table III with degrees of freedom equal to infinity, you calculate

$$38.333 \pm (1.96)(3.012 / \sqrt{499}) = 38.333 \pm .26428 = (38.07, 38.60) \text{ weeks.}$$

You are 95% certain the average number of weeks of gestation for all babies born in North Carolina in 2005 was between 38.07 and 38.60 weeks.

Number of prenatal doctor visits:

a. The graph of the number of prenatal doctor visits is almost normal, although there are some high outliers and a slightly skew to the left. The number of visits ranges from a low of no visits to a high of 30 visits (this was for a pregnancy that lasted 37 weeks, so there was almost one doctor visit a week). The typical number of prenatal doctor visits in this sample was 12, and half of the mother's made between 10 and 15 prenatal visits to a doctor.

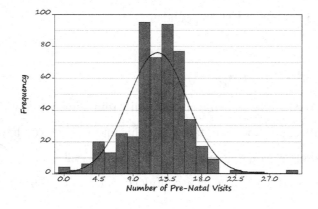

b. See above.

c. Using Table III with degrees of freedom equal to infinity, you calculate

$$12.207 \pm (1.96)(3.922 / \sqrt{498}) = 12.207 \pm .34447 = (11.86, 12.55) \text{ visits.}$$

You are 95% certain the average number of prenatal doctor visits for all babies born in North Carolina in 2005 was between 11.86 and 12.55 visits per child.

Weight gained by mother:

a. The weights gained by the mothers are fairly symmetric with a minimum of 0 pounds (!?) to a maximum of 65, with high outliers at 70, 72 and 75 pounds. Fifteen of these 497 mothers gained no weight during their pregnancies, but most of the mothers gained between 20 and 40 pounds. The typical weight gain for these mothers was 30 pounds.

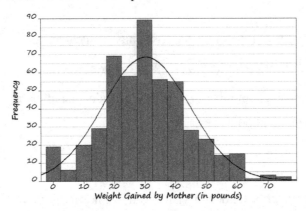

b. See above.

c. Using Table III with degrees of freedom equal to infinity, you calculate

$$30.4 \pm (1.96)(14.177/\sqrt{497}) = 30.4 \pm 1.2464 = (29.15, 31.65) \text{ pounds.}$$

You are 95% certain the average weight gained by the mothers of all babies born in North Carolina in 2005 was between 29.15 pounds and 31.65 pounds.

Exercise 19-31: M&M Consumption

a. You should not expect 95% of the *sample* values to fall within the 95% confidence interval. You would expect 95% of all sample means to fall within this interval, but you have no idea how many of the sample values themselves would fall in any confidence interval.

b. Nine of the sample values fall within the 95% confidence interval (35.96, 64.34) candies. This is 45% of the sample values.

c. You would not expect 95% of the *population* values to fall within the 95% confidence interval either. The confidence interval should contain the population mean value, but you have no idea what percentage of the population values would lie in this interval.

Topic 20

Tests of Significance: Means

Odd- Numbered Exercise Solutions

Exercise 20-5: Exploring the *t*-Distribution

a. Here is the completed table:

df	Pr(T ≥ 1.415)	Pr(T ≥ 1.960)	Pr(T ≥ 2.517)	Pr(T ≥ 3.168
4	.1 < p-value < .2	.05 < p-value < .1	.025 < p-value < .05	.01 < p-value < .025
11	.05 < p-value < .1	.025 < p-value < .05	.01 < p-value < .025	.001 < p-value < .005
23	.05 < p-value < .1	.025 < p-value < .05	.005 < p-value < .01	.001 < p-value < .005
80	.05 < p-value < .1	.025 < p-value < .05	.005 < p-value < .01	.001 < p-value < .005
Infinity	.05 < p-value < .1	p-value = .025	.005 < p-value < .01	.0005 < p-value < .001

b. As the value of the test statistic increases, the *p*-value gets smaller.

c. As the number of degrees of freedom increases, the *p*-value gets smaller.

Exercise 20-7: Sleeping Times

H_0: $\mu = 7.0$ H_a: $\mu < 7.0$

Sample 1: test statistic $t \approx -1.53$;   *p*-value = 0.080

Sample 2: test statistic $t \approx -0.79$;   *p*-value = 0.224

Sample 3: test statistic $t \approx -2.66$;   *p*-value = 0.006

Sample 4: test statistic $t \approx -1.37$;   *p*-value = 0.090

At the 5% significance level, you would reject H_0 and conclude the population mean sleep time is less than 7 hours only for Sample 3 (*p*-value = .006).

This is the same result you had with the two-sided test. For each sample, the test statistic value was the same with the two-sided test, but the *p*-value was twice as large.

Exercise 20-9: Basketball Scoring

a. The null hypothesis is that the mean number of points scored per game for the entire 1999-2000 season is 183.2. In symbols, $H_0: \mu = 183.2$.

The alternative hypothesis is that the mean number of points scored per game for the entire 1999-2000 season is greater than 183.2. In symbols, $H_a: \mu > 183.2$.

b. No, you do not have enough information to calculate the test statistic. You need to know the standard deviation of the first 149 games.

c.   Using Table III and 100 degrees of freedom, the test statistic would need to be at least $t = 2.364$ in order to reject the null hypothesis. Using Minitab, the test statistic would need to be $t = 2.35281$.

d. If you assume the standard deviation was about $s = 20.27$, then $t = \dfrac{196.2 - 183.2}{20.27 / \sqrt{149}} \approx 7.83$, which does exceed the rejection value in part c by a great deal.

e. Yes. Even though the magazine did not provide the standard deviation, using a reasonable value for the standard deviation would result in a test statistic that is significantly greater than the value needed to reject the null hypothesis at the .01 significance level. So, you can reasonably predict that the results (reject H_0 and conclude the mean number of points per game is greater than 183.2) would be statistically significant at the .01 level.

f. No, the validity of this test procedure does not depend on the scores being normally distributed because the sample size is large (149 > 30). (We do still need to worry about these games not being a random sample however.)

Exercise 20-11: Body Temperatures

The following output pertains to the data for female and male body temperatures.

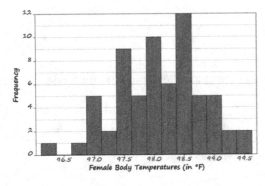

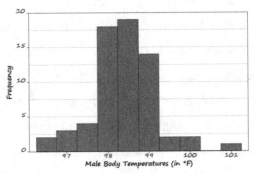

	n	Mean	SD	Min.	Q_L	Median	Q_U	Max.
Females	65	98.105	0.699	96.3	97.6	98.1	98.6	99.5
Males	65	98.394	0.743	96.4	98.0	98.4	98.8	100.8

Females:

Define parameter of interest: Let μ represent the average adult female body temperature.

The null hypothesis is that the average adult female body temperature is 98.6°F. In symbols, the null hypothesis is H_0: $\mu = 98.6$°F.

The alternative hypothesis is that the average adult female body temperature is not 98.5°F. In symbols, the alternative hypothesis is H_a: $\neq 98.6$°F.

Check technical conditions: The sample size is large ($n = 65 > 30$), but you do not know whether this sample was randomly selected (in fact, they were volunteers in another clinical trial, so this sample may not be representative of healthy adults).

Test statistic is $t = \dfrac{98.105 - 98.6}{0.699 / \sqrt{65}} \approx -5.71$.

Using Table III with 60 degrees of freedom, p-value $< 2 \times .0005 = .001$.
Using Minitab, p-value $= 2 \times .0000002 = .0000004$.

Test decision: With such a small p-value, reject H_0 at any commonly used significance level.

Conclusion in context: You have very strong statistical evidence (p-value $< .001$) that the average adult female body temperature is not 98.6°F.

Males:

Define parameter of interest: Let μ represent the average adult male body temperature.

The null hypothesis is that the average adult male body temperature is 98.6°F. In symbols, the null hypothesis is H_0: $\mu = 98.6$°F.

The alternative hypothesis is that the average adult male body temperature is not 98.5°F. In symbols, the alternative hypothesis is H_a: $\neq 98.6$°F.

Check technical conditions: The sample size is large ($n = 65 > 30$), but you do not know whether this sample was randomly selected (in fact, they were volunteers in another clinical trial, so this sample may not be representative of healthy adults).

Test statistic is $t = \dfrac{98.394 - 98.6}{0.743 / \sqrt{65}} \approx -2.24$.

Using Table III with 60 degrees of freedom, $2 \times .01 < p\text{-value} < 2 \times .025$, or $.02 < p\text{-value} < .05$. Using Minitab, $p\text{-value} = 2 \times .0142856 = .02857$.

Test decision: With a $p\text{-value} = .029 < .05$, reject H_0 at the .05 significance level.

Conclusion in context: You have moderate statistical evidence ($p\text{-value} - .029$) that the average adult male body temperature is not 98.6°F.

Although you reach the same conclusion for males and females, the evidence is much stronger in the case of the females (as evidenced by the smaller $p\text{-value}$).

Exercise 20-13: Age Guesses

Answers will vary. Here is one representative set of answers:

The following output describes the age guesses:

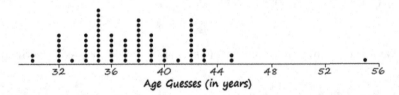

Age Guesses (in years)

Variable	n	Mean	SD	Min.	Q_L	Median	Q_U	Max.
Age guesses	71	37.532	4.157	30	35	37	40	55

Define parameter of interest: Let μ represent the mean guess of this instructor's age by all students at this school.

The null hypothesis is that the population mean guess of this instructor's age was correct (40 years). In symbols, the null hypothesis is H_0: $\mu = 40$.

The alternative hypothesis is that population mean guess of this instructor's age was not correct. In symbols, the alternative hypothesis is H_a: $\mu \neq 40$.

Check technical conditions: This was not a random sample but it may be representative of the guesses for this instructor's age that students would typically make at this school. You do not need to worry about the bias of under-guessing by students in order not to offend the instructor. The sample size is large (71 > 30), so this condition is met for this sample.

Test statistic is $.t = \dfrac{37.521 - 40}{4.157 / \sqrt{71}} \approx -5.02$

Using Table III with 60 degrees of freedom, p-value < 2 × .0005 = .001. Using Minitab, p-value = 2 × .0000019 = .0000038.

Test decision: With such a small p-value, reject H_0 at any common significance level.

Conclusion in context: You have very strong statistical evidence (p-value < .001) that the population mean guess of this instructor's age was not correct (was not 40 years), as long as the guesses made by this sample are representative.

Exercise 20-15: Nicotine Lozenge

a. The null hypothesis is that the population mean number of cigarettes smoked per day is 20. In symbols, H_0: $\mu = 20$.

The alternative hypothesis is that the population mean number of cigarettes smoked per day is not 20. In symbols, H_a: $\mu \neq 20$.

Technical conditions: The sample size is large (1818 > 30), but the subjects were not randomly selected.

The test statistic is $t = \dfrac{22 - 20}{10.8 / \sqrt{1818}} \approx 7.90$.

Using Table III with degrees of freedom equal to infinity, p-value < 2 × .0005 = .001. Using Minitab, the p-value is .000001.

Because of the small p-value, reject H_0 at any common significance level.

If this sample is representative of the general population, you have very strong statistical evidence (p-value $< .001$) that the population mean number of cigarettes smokes per day is not 20 (one pack).

b. The null hypothesis is that the population mean number of cigarettes smoked per day is 20. In symbols, H_0: $\mu = 20$.

The alternative hypothesis is that the population mean number of cigarettes smoked per day is not 20. In symbols, H_a: $\mu \neq 20$.

The test statistic is $t = \dfrac{22 - 20}{10.8 / \sqrt{100}} \approx 1.85$.

Using Table III with 80 degrees of freedom, $2 \times .025 < p\text{-value} < 2 \times .05$, so $.05 < p\text{-value} < .10$. Using Minitab, p-value $= 2 \times .0336481 = .0673$.

Barely fail to reject H_0 at the $\alpha = .05$ significance level.

You do not have sufficient statistical evidence (at the 5% level) to conclude the population mean number of cigarettes smoker per day differs from 20.

c. You reached different conclusions in parts a and b; rejecting the null hypothesis when the sample size was very large, but failing to reject it when the sample size was only 100. This result makes sense because with larger sample sizes there is less random sampling variability and the same sample mean will be more surprising/less likely to happen by chance alone.

Exercise 20-17: Backpack Weights

a. This is a quantitative variable.

b. The null hypothesis is that the population mean ratio of backpack weight to body weight is equal to .10. In symbols, the null hypothesis is H_0: $\mu = .10$.

The alternative hypothesis is that the population mean ratio of backpack weight to body weight is not equal to .01. In symbols, the alternative hypothesis is H_a: $\mu \neq .10$.

Technical conditions: The sample was not randomly selected, but the researchers did try to select a representative sample. The sample size is large ($n = 100 > 30$). You may consider the technical conditions to have been met.

The test statistic is $t = \dfrac{.0771 - .1}{.0366 / \sqrt{100}} \approx -6.26$.

Using Table III with 80 degrees of freedom, p-value $< 2 \times .0005 = .0010$.

Using Minitab, p-value $\approx .0000$.

Because the p-value is small, reject H_0 at any reasonable significance level.

You have very strong statistical evidence (p-value $< .001$) that the population mean ratio is not .10.

c. The 99% confidence interval given in Activity 19-6 is (.0674, .0868). Note that .10 is not in this interval, which implies that you would reject this as a plausible value for μ at the $\alpha = .01$ significance level. This is consistent with your test results.

Exercise 20-19: Nicotine Lozenge

a. From Exercise 19-11, a 99% confidence interval for μ is (21.3475, 22.6525). Thus, any value in this interval is a plausible value for μ, and any value *not* in this interval would be rejected at the $\alpha = .01$ significance level.

b. The null hypothesis is that the population mean number of cigarettes smoked per day by all subjects who might be in such a study is 22. In symbols, the null hypothesis is H_0: $\mu = 22$.

The alternative hypothesis is that the population mean number of cigarettes smoked per day is not 22. In symbols, the alternative hypothesis is H_a: $\mu \neq .22$.

Technical conditions: The sample size is large ($n = 1818 > 30$), but the subjects were not randomly selected, so proceed with caution.

The test statistic is $t = \dfrac{22 - 22}{10.8 / \sqrt{1818}} = 0$.

Using Table III with degrees of freedom equal to infinity, p-value $> 2 \times .1 = .4$.

Using Minitab, p-value $= 2 \times .5 = 1.0$.

With this large p-value is small, do not reject H_0 at the $\alpha = .05$ significance level.

You do not have sufficient evidence to conclude that the population mean number of cigarettes smoked per day differs from 22.

c. Based on this p-value, 22 would be in the 95% confidence interval for μ (in fact, because it equals the sample mean, it would be at the center of the interval).

Exercise 20-21: Pet Ownership

a. The null hypothesis is that the mean number of cats in all American cat-owning households is 2.0. In symbols, H_0: $\mu = 2.0$.

The alternative hypothesis is that the mean number of cats in all American cat-owning households is more than 2.0. In symbols, H_a: $\mu > 2.0$.

b. You need to know the standard deviation.

c. Answers will vary according to the chosen value of s.

Assuming s is 1.0, the test statistic is $t = \dfrac{2.1 - 2}{1/\sqrt{25,280}} \approx 15.9$.

Using Table III with degrees of freedom equal to infinity, p-value $<.0005$.

Using Minitab, p-value $\approx .000000$.

With this small p-value, reject H_0 at any reasonable significance level.

You have very strong statistical evidence that the mean number of cats per cat-owning household is greater than 2.0.

d. Now assuming $s = 2.0$, the test statistic is $t = \dfrac{2.1 - 2}{2/\sqrt{25,280}} \approx 7.95$.

Using Table III with degrees of freedom equal to infinity, p-value $<.0005$.

Using Minitab, p-value $\approx .000000$.

With this small p-value, reject H_0 at any reasonable significance level.

You have very strong statistical evidence that the mean number of cats per cat-owning household is greater than 2.0. Your conclusion did not change.

e. For a 99% CI for μ with degrees of freedom equal to infinity, you calculate $2.1 \pm (2.576)(.00629) = (2.0838, 2.1162)$ cats per household. Clearly, the population mean does not exceed 2.0 in any practical sense as you are 99% sure the population mean is between 2.084 and 2.116 cats per household.

Exercise 20-23: Birth Weights

a. Let μ be the mean birth weight of the population of babies born in North Carolina in 2005.

The null hypothesis is that the population mean birth weight is 7.275 pounds. In symbols, the null hypothesis is H_0: $\mu = 7.275$.

The alternative hypothesis is that the population mean birth weight is not 7.275 pounds. In symbols, the alternative hypothesis is H_a: $\mu \neq 7.275$.

Technical conditions: The sample was randomly selected and the sample size is large ($n = 500 > 30$), so the technical conditions necessary for the validity of this test procedure are met.

The test statistic is $t = \dfrac{7.0687 - 7.275}{1.5062 / \sqrt{500}} \approx -3.06$.

Using Table III with degrees of freedom equal to infinity, $2 \times .001 < p$-value $< 2 \times .005$, so $.002 < p$-value $< .01$.

Using Minitab, p-value $= 2 \times .0011662 = .0023324$.

Because the p-value is so small, reject H_0 at any reasonable significance level.

You have found strong statistical evidence that the population mean birth weight of all babies born in North Carolina in 2005 is not 7.275 pounds.

Exercise 20-25: Birth Weights

a. Because the value $\mu_0 = 30$ is contained in all three of the given confidence intervals, you know that the two-sided p-value for testing H_0: $\mu = 30$ vs. H_a: $\mu \neq 30$ is greater than .10.

b. Because p-value $> .10$, you would fail to reject H_0 at the .10, .05 and .01 significance levels, and conclude that there is not convincing evidence that the population mean age of all North Carolina fathers in 2005 differs from 30 years.

Exercise 20-27: Basketball Scoring

a. If the rule change had no effect on scoring, the null hypothesis would be H_0: $\mu = 183.2$.

b. If the rule change had the desired effect, the alternative hypothesis would be H_a: $\mu > 183.2$.

c. Here is a dotplot of the data:

Points per Game (Dec. 10-12, 1999)

This dotplot does seem to indicate that the average number of points scored per game has increased. Among these twenty-five games there was one extremely low outlier in which only 140 points were scored, and another in which 168 points were scored. In all of the remaining games, at least 180 points were scored. The mean number of points per game was 195.88, and the standard deviation was 20.27. The maximum number of points scored was 243.

d. $\bar{x} = 195.88$ points per game; $s = 20.27$ points per game

e. The sample was not randomly selected from all of the NBA games during the 1999-2000 season; in fact all of these games were played early in the season during one particular three-day period. Therefore this is probably not a representative sample of the points scored per game during this season. In addition, the sample size is small ($n = 25 < 30$) and the data do not appear to be normally distributed. Thus the technical conditions necessary for the validity of this test do not appear to be satisfied, so you should proceed with caution.

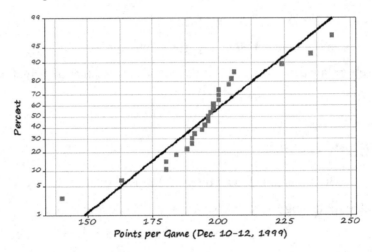

f. The test statistic is $t = \dfrac{195.88 - 183.2}{20.27 / \sqrt{25}} = 3.13$.

g. Using Table III with 24 degrees of freedom, $.001 < p\text{-value} < .005$.

h. This p-value indicates that if the rule change had no effect on scoring, you would see a sample mean of 195.88 or more points in a random sample of 25 games from this season between .1% and .5% of the time.

i. Because $.001 < p\text{-value} < .005$, you would reject the null hypothesis at the .10, .05, and .01 levels, but you would not reject H_0 at the .001 level.

j. Because this sample result (an average of 195.88 or more points per game in a sample of 25 games) would occur less than .5% of the time by random chance alone if the rule change had no effect on scoring, you have found very strong statistical evidence that average number of points per game during the 1999–2000 season is now more than 183.2 points. However, this sample was not randomly selected from all NBA games during the season, so it may not be representative of whether the scoring changed over the entire season. Note that the t-procedures are fairly robust against deviations from normality as long as there are no extreme outliers, but there is a fairly low outlier, so you should be cautious about putting too much weight into the results of this test.

Unit 5

Inference from Data: Comparisons

Topic 21

Comparing Two Proportions

Odd- Numbered Exercise Solutions

Exercise 21-7: Botox for Back Pain

a. The explanatory variable is *whether the subject received Botox or the placebo* (ordinary saline injection).

The response variable is *whether the subject experiences substantial pain relief.*

b. Here is the 2 × 2 table:

	Botox	Saline	Total
Substantial Pain Relief	9	2	11
No Pain Relief	6	14	20
Total	15	16	31

c. Based on the histogram, the approximate *p*-value is 7/1000, or .007. This value was calculated by approximating the number of times nine or more successes were assigned to the Botox group.

d. This *p*-value means that if Botox and saline are equally effective in treating low-back pain, then you would expect to see such sample results (9 or more of 11 successes assigned to a group of 15) about .7% of the time by random assignment alone. This occurrence is very rare, however, so you would conclude that there is very strong statistical evidence that Botox is more effective than saline in treating low-back pain.

e. Technical conditions:

   i. The data were obtained by randomly assigning subjects to treatment groups. This condition is met.

   ii. Checking $n_1\hat{p}_c \geq 10$ with $\hat{p}_c = .3548$, and $n_1 = 15$, $n_2 = 16$, we find $n_1\hat{p}_c = 5.322$ and $n_2\hat{p}_c = 5.677$, so the sample sizes are not quite large for this condition to be met.

Exercise 21-9: Perceptions of Self-Attractiveness

a. For $n_1 = n_2 = 100$: $(.81 - .71) \pm (1.96) \sqrt{\dfrac{(.81)(.19)}{100} + \dfrac{(.71)(.29)}{100}} = (-.0176, .2176)$

For $n_1 = n_2 = 200$: $(.81 - .71) \pm (1.96) \sqrt{\dfrac{(.81)(.19)}{200} + \dfrac{(.71)(.29)}{200}} = (.0169, .1831)$

For $n_1 = n_2 = 500$: $(.81 - .71) \pm (1.96) \sqrt{\dfrac{(.81)(.19)}{500} + \dfrac{(.71)(.29)}{500}} = (.0474, .1526)$

b. For $n_1 = n_2 = 100$, half-width = .117567.

　　For $n_1 = n_2 = 200$, half-width = .083133.

　　For $n_1 = n_2 = 500$, half-width = .052578.

　　The half-width decreases as the sample size increases. This is consistent with what happens with confidence intervals for a single proportion.

c. With sample sizes $n_1 = n_2 = 100$, the 95% confidence interval includes the value zero. With sample sizes $n_1 = n_2 = 200$ and $n_1 = n_2 = 500$, the confidence intervals do not include zero. This is consistent with your test results in Activity 21-3, where you found that the test H_0: $\pi_m = \pi_f$ vs. H_a: $\pi_m \neq \pi_f$ was statistically significant when $n_1 = n_2 = 200$ and $n_1 = n_2 = 500$, but not when $n_1 = n_2 = 100$.

Exercise 21-11: Generation M

a. $\pi_g - \pi_b$ represents the difference in the population proportion of girls who have televisions in their bedroom and the population proportion of boys who have televisions in their bedrooms.

b. You calculate $(.64 - .72) \pm (1.96) \sqrt{\dfrac{(.64)(.36)}{1036} + \dfrac{(.72)(.28)}{996}} = -.08 \pm (1.96)(.0206) = (-.1204, -.0396)$.

　　You are 95% confident that the proportion of all girls who have televisions in their bedrooms is somewhere between .0396 and .1204 less than the proportion of all boys who have televisions in their bedrooms. The fact that all the values in your confidence interval are negative indicates that the population proportion of girls (with televisions in their bedrooms) is strictly less than the population proportion of boys.

c. For a 99% confidence interval you calculate $(.64 - .72) \pm (2.576)(.0206) = (-.1331, -.0269)$. The midpoint of both intervals is the same $(.64 - .72) = -.08$. The 99% confidence interval is wider than the 95% confidence interval.

Exercise 21-13: AZT and HIV

a. The following segmented bar graph displays the results:

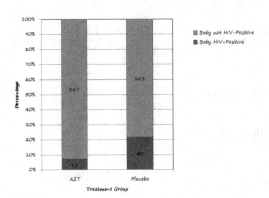

This bar graph indicates that mothers given the placebo were about three times more likely to have babies that were HIV positive than were the mothers given AZT.

b. Technical conditions:

i. The data are from randomly assigning subjects to two treatment groups, so this condition is met.

ii. The number of successes and failures in each group should be at least ten. This condition is also met as the smallest of these counts is 13.

c. The null hypothesis is that AZT and a placebo are equally effective in reducing mother-to-infant transmissions of AIDS. Specifically, the proportion of HIV-positive babies born to mothers who could potentially take AZT is the same as the proportion of HIV-positive babies born to mothers who could potentially take a placebo. In symbols, the null hypothesis is H_0: $\pi_{AZT} = \pi_{placebo}$.

The alternative hypothesis is that AZT is more effective than a placebo for reducing mother-to-infant transmissions of AIDS, or that the proportion of HIV-positive babies born to mothers who could potentially take AZT is smaller than the proportion of HIV-positive babies born to mothers who could potentially take a placebo. In symbols, the alternative hypothesis is H_a: $\pi_{AZT} < \pi_{placebo}$.

Test statistic: $z = \dfrac{.0722 - .2186}{\sqrt{(.146)(1-.146)\left(\dfrac{1}{180} + \dfrac{1}{183}\right)}} = -3.95$

p-value = $\Pr(Z < -3.95) \le .0002$.

With such a small p-value, reject H_0 at the $\alpha = .01$ significance level.

You have very strong evidence that AZT is more effective than a placebo in reducing mother-to-infant transmission of AIDS.

d. For a 99% confidence interval, you calculate

$$(.0722 - .2186) \pm (2.576)\sqrt{\dfrac{(.0722)(.9278)}{180} + \dfrac{(.2186)(.7814)}{183}} = (-.2395, -.0533).$$

You are 99% confident the difference in HIV transmission rates is between 5.33 and 23.95 percentage points. Because the values in your interval are all negative, you know that the AZT transmission rate is lower than the placebo transmission rate by somewhere between 5.33 and 23.95 percentage points.

e. Because this is a well-designed experiment, you can conclude that AZT caused the observed difference in HIV transmission rates. If AZT and a placebo were equally effective in reducing mother-to-infant transmission of AIDS, you would virtually never see a difference in sample results as or more extreme as those seen in this experiment by random assignment alone. You are 99% confident in concluding that AZT lowers the HIV transmission rate somewhere between 5.33 and 23.95 percentage points over a placebo.

Exercise 21-15: Suitability for Politics

a. The null hypothesis is that the proportion of all American adult men who agree with this statement is the same as the proportion of all American adult women who agree with this statement. In symbols, H_0: $\pi_M = \pi_W$.

The alternative hypothesis is that the proportion of all American adult men who agree with this statement is greater than the proportion of all American adult women who agree with this statement. In symbols, H_a: $\pi_M > \pi_W$.

b. Technical conditions: Using $\hat{p}_c = .254$, we find $n_1\hat{p}_c = 109$, $n_1(1-\hat{p}_c) = 276$, $n_2\hat{p}_c = 103$, and $n_2(1-\hat{p}_c) = 346$, all of which are greater than 10, so this condition is met. However, the data are technically not independent random samples from two populations – this is *one* random sample taken from a population and then separated by gender. However, you might consider this a minor variation and say the technical condition has been met.

c. The test statistic is $z = \dfrac{.283 - .229}{\sqrt{(.254)(1-.254)\left(\dfrac{1}{385} + \dfrac{1}{449}\right)}} = 1.79$.

Using Table II, p-value = $\Pr(Z > 1.79) = .0367$.

Because the p-value = $.0367 < .05$, reject H_0 at the $\alpha = .05$ significance level.

You have moderate statistical evidence that the proportion of American adult men who agree with this statement is greater than the proportion of American adult women who agree with this statement.

d. If the population proportions of men and women who agree with this statement were the same, you would see sample results as or more extreme as this (a difference of at least .054 with these sample sizes) in about 3.67% of samples by random sampling alone. Because this would not be a very common occurrence, these results provide moderate evidence that the assumed hypothesis (the proportions of men and women who agree with this statement are the same) is false. You have moderate statistical evidence that the population proportion of men who agree with this statement is actually greater than the population proportion of women who agree.

e. For a 90% CI, you calculate $(.283 - .229) \pm (1.645)\sqrt{\dfrac{(.283)(.717)}{385} + \dfrac{(.229)(.771)}{449}} = (.0041, .1039)$.

f. You are 90% confident the percentage of all adult American men who agree with this statement is between .4 and 10.9 percentage points *higher than* the percentage of all adult American women who agree with this statement.

Exercise 21-17: Volunteerism

a. These are statistics because they are taken from samples, not populations.

b. You need to know how many men were sampled and how many women.

c. For a 99% CI, you calculate $(.25 - .324) \pm (2.576)\sqrt{\dfrac{(.25)(.75)}{30000} + \dfrac{(.324)(.676)}{30000}} = (-0.0835, -0.0645)$.

You are 99% confident the percentage of American men who did volunteer work in 2005 was between 6.45 and 8.35 percentage points less than the percentage of American women who did volunteer work in 2005.

d. This interval is so narrow because the sample sizes are so large (30,000 each).

e. A 99.9% confidence interval should be wider because to create it you would need to use a larger z^* (3.291 rather than 2.576).

Exercise 21-19: Underhanded Free Throws

a. The null hypothesis is that Reilly's probability of success is the same with both methods of shooting free throws. In symbols, H_0: $\pi_{underhand} = \pi_{conventional}$.

The alternative hypothesis is that Reilly's probability of success is greater with the underhand method than with the conventional method. In symbols, H_a: $\pi_{underhand} > \pi_{conventional}$.

b. You would need to know how many attempts he made with each method.

c. Answers will vary by student, but students should expect the results to be more significant if he made 500 attempts using each method than if he made 100 attempts using each method.

d. The test statistic is $z = \dfrac{.78 - .63}{\sqrt{(.705)(1-.705)\left(\dfrac{1}{100} + \dfrac{1}{100}\right)}} = 2.33$.

Using Table II, p-value = (Z > 2.33) = .0099.

Reject H_0 at the $\alpha = .01$ significance level because the p-value = .0099 < .01.

e. The test statistic is $z = \dfrac{.78 - .63}{\sqrt{(.705)(1-.705)\left(\dfrac{1}{500} + \dfrac{1}{500}\right)}} = 5.20$.

Using Table II, p-value = $\Pr(Z > 5.20) \approx .0000$.

Reject H_0 at the $\alpha = .01$ significance level because p-value < .01.

f. In both cases, you would reject the null hypothesis and conclude that you have very strong statistical evidence that Reilly's proportion of successes is greater with the underhand method than with the conventional method. Note that as predicted, the results are more statistically significant with the larger sample size.

g. No, you would not feel comfortable generalizing these results to all players who could be taught by Rick Barry. These results were not obtained from a random sample of players who were taught by Rick Barry. Perhaps both Reilly and Barry are right-handed or have some other common feature that makes the underhand method work for them, but not for others.

h. A Type I error would be concluding that Reilly's probability of success is greater with the underhand method than with the conventional method, when both methods are actually equally effective. If he makes this type of error, the consequences would be minimal; he would tend to shoot free throws using the underhand method, when he could be using either method.

A Type II error would be failing to realize that Reilly's probability of success is greater with the underhand method than with the conventional method when it really is. This type of error would probably mean that Reilly does not succeed in his free throwing as often as he might because he would not primarily use the underhand method.

Exercise 21-21: Magazine Advertisements

a. $\hat{p}_{SI} = .466$; $\hat{p}_{SOD} = .215$.

b. The following segmented bar graph compares the proportions:

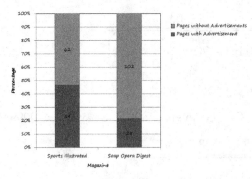

c. The null hypothesis is that the proportion of all pages with ads in both magazines is the same. In symbols, H_0: $\pi_{\text{Sports Illustrated}} = \pi_{\text{Soap Opera Digest}}$.

The alternative hypothesis is that the proportion of all pages with ads in both magazines is not the same. In symbols, H_a: $\pi_{\text{Sports Illustrated}} \neq \pi_{\text{Soap Opera Digest}}$.

The test statistic is $z = \dfrac{.466 - .215}{\sqrt{(.333)(1 - .333)\left(\dfrac{1}{116} + \dfrac{1}{130}\right)}} = 4.15$

p-value $= 2 \times \Pr(Z > 4.15) \approx 2 \times .0000 = .0000$.

d. Because the p-value is less than $\alpha = .01$, reject H_0 and conclude that there is very strong statistical evidence that the proportion of pages with ads in both magazines is not the same. (If they were the same, you would virtually never see a sample difference this extreme or more extreme by random sampling alone.)

e. In order for this test procedure to be valid, you must assume that these are independent random samples of pages from the magazines.

Exercise 21-23: Wording of Surveys

a. The sample proportions are $\hat{p}_{forbid} = 161/409 = .393$ and $\hat{p}_{\text{allow}} = (432 - 189)/432 = .5625$.

b. The null hypothesis is that the population proportion of all potential "forbid" subjects who oppose communist speeches is the same as the population proportion of potential "allow" subjects who would allow communist speeches. In symbols, H_0: $\pi_{\text{Forbid}} = \pi_{\text{Allow}}$.

The alternative hypothesis is that the population proportion of all potential "forbid" subjects who oppose communist speeches is not the same as the population proportion of potential "allow" subjects. In symbols, H_a: $\pi_{\text{Forbid}} \neq \pi_{\text{Allow}}$.

Technical conditions: The values $n_1\hat{p}_c$, $n_1(1-\hat{p}_c)$, $n_2\hat{p}_c$, and $n_2(1-\hat{p}_c)$ are all greater than 10 (the smallest is 191), and the subjects were presumably randomly assigned to the different question version.

The test statistic is $z = \dfrac{.39 - .5625}{\sqrt{(.480)(1-.480)\left(\dfrac{1}{409} + \dfrac{1}{432}\right)}} = -4.90$.

p-value $= 2 \times \Pr(Z < -4.90) \approx 2 \times .0000 = .0000$.

Reject H_0 at the $\alpha = .10$, $.05$ and $.01$ significance level because the p-value $< .0001$.

You have sufficient statistic evidence to conclude that the population of subjects who oppose communist speeches is affected by the use of the words "forbid" and "allow" in the question.

a. The null hypothesis is that the population proportion of all potential "forbid" subjects who oppose X-rated movies is the same as the population proportion of potential "allow" subjects who would oppose X-rated movies. In symbols, H_0: $\pi_{Forbid} = \pi_{Allow}$.

The alternative hypothesis is that the population proportion of all potential "forbid" subjects who oppose X-rated movies is not the same as the population proportion of potential "allow" subjects. In symbols, H_a: $\pi_{Forbid} \neq \pi_{Allow}$.

The sample proportions are $\hat{p}_{Forbid} = .4095$ and $\hat{p}_{Allow} = .4635$.

Technical conditions: The values $n_1\hat{p}_c$, $n_1(1-\hat{p}_c)$, $n_2\hat{p}_c$, and $n_2(1-\hat{p}_c)$ are all greater than 10 (the smallest is 224), and the subjects were presumably randomly assigned to the different question version.

The test statistic is $z = \dfrac{.4095 - .4635}{\sqrt{(.4372)(1-.4372)\left(\dfrac{1}{547} + \dfrac{1}{576}\right)}} = -1.82$.

p-value $= 2 \times \Pr(Z < -1.82) \approx 2 \times .0344 = .0688$.

Reject H_0 at the $\alpha = .10$ significance level, but do not reject H_0 at the $\alpha = .05$ or $.01$ significance levels because $.05 < p$-value $< .10$.

b. The null hypothesis is that the population proportion of all potential "forbid" subjects who oppose cigarette ads on television is the same as the population proportion of potential "allow" subjects. In symbols, H_0: $\pi_{Forbid} = \pi_{Allow}$.

The alternative hypothesis is that the population proportion of all potential "forbid" subjects who oppose cigarette ads on television is not the same as the population proportion of potential "allow" subjects. In symbols, H_a: $\pi_{Forbid} \neq \pi_{Allow}$.

The sample proportions are $\hat{p}_{forbid} = .506$ and $\hat{p}_{allow} = .764$.

Technical conditions: The values $n_1\hat{p}_c$, $n_1(1-\hat{p}_c)$, $n_2\hat{p}_c$, and $n_2(1-\hat{p}_c)$ are all greater than 10 (the smallest is 134), and the subjects were presumably randomly assigned to the different question version.

The test statistic is $z = \dfrac{.506 - .764}{\sqrt{(.63)(1-.63)\left(\dfrac{1}{607} + \dfrac{1}{576}\right)}} = -9.15$.

p-value $= 2 \times \Pr(Z < -9.15) \approx 2 \times .0000 = .0000$.

Reject H_0 at the $\alpha = .10$, .05 and .01 significance level because the p-value $< .0001$.

c. It appears that, in general, the words "forbid" and "allow" are not interchangeable in survey questions, particularly regarding social values issues. They made a very (statistically) significant difference in the answers to the questions about cigarette ads on television and in the answers to the question about communist speeches, and a moderately statistically significant difference in the answers to the X-rated movies question.

Exercise 21-25: Teen Smoking

a. Here is the segmented bar graph:

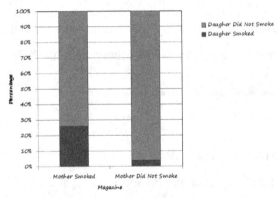

b. This is an observational study because the researchers did not determine which mothers would or would not smoke during their pregnancies. This explanatory variable was determined by the mothers themselves.

c. You would need to know how many girls were interviewed in order to determine whether the difference in sample proportions is statistically significant.

d. The null hypothesis is that daughters of mothers who smoked during their pregnancies are just as likely to smoke themselves as daughters of mothers who did not smoke during their pregnancies. In symbols, H_0: $\pi_{\text{mother smoked}} = \pi_{\text{mother didn't smoke}}$.

The alternative hypothesis is that daughters of women who smoked during pregnancy are more likely to smoke themselves than are daughters of women who did not smoke during pregnancy. In symbols, $H_a: \pi_{\text{mother smoked}} > \pi_{\text{mother didn't smoke}}$.

Technical conditions: Using a combined proportion $\hat{p}_c = .15 = (2 + 13)/100$, you find that $n_1\hat{p}_c = 15 \times .04 = 7.5$, which is not at least 10. So the sample size condition is not met. In addition, you are given no information about how the sample of teenagers was selected, so you should be extremely cautious about generalizing these results beyond this particular sample.

The test statistic is $z = \dfrac{.26 - .04}{\sqrt{(.15)(1-.15)\left(\dfrac{1}{50}+\dfrac{1}{50}\right)}} = 3.08$.

p-value $= 2 \times \Pr(Z > 3.08) = .0010$.

With the p-value $= .001 < .05$, reject H_0 at the $\alpha = .05$ significance level.

You conclude there is strong statistical evidence that daughters of mothers who smoke during pregnancy are more likely to smoke themselves than daughters of mothers who do not smoke during pregnancy.

e. Because this is an observational study and not an experiment, although the results are statistically significant, you cannot conclude that the pregnant mothers' smoking causes the daughters' tendency to smoke. Possible confounding variables include whether the mothers continued to smoke through the daughter's youth and/or whether their fathers or some other household member smoked during their childhood (assuming this is more likely to be true for those whose mother smoked during pregnancy).

f. A Type I error would be concluding that daughters of mothers who smoked during their pregnancies are more likely to smoke themselves as daughters of mothers who did not smoke during their pregnancies, when they are actually no more likely to smoke themselves.

A Type II error would be failing to recognize that daughters of mothers who smoked during their pregnancies are more likely to smoke themselves as daughters of mothers who did not smoke during their pregnancies when they actually are.

Exercise 21-27: Candy and Longevity

a. The null hypothesis is that the population proportion of all candy consumers who die (during this period) is the same as the population proportion of all non-consumers who die. In symbols, $H_0: \pi_{\text{candy}} = \pi_{\text{non-consumer}}$.

The alternative hypothesis is that the population proportion of all candy consumers who die (during this period) is less than the population proportion of all non-consumers who die. In symbols, $H_a: \pi_{candy}$ $< \pi_{non-consumer}$.

The sample proportions are $\hat{p}_{candy} = .059$ and $\hat{p}_{non-consumer} = .075$.

The test statistic is $z = \dfrac{.059 - .075}{\sqrt{(.066)(1 - .066)\left(\dfrac{1}{4529} + \dfrac{1}{3312}\right)}} = -2.82$.

p-value $= \Pr(Z < -2.82) = .0024$.

With p-value $= .0024 < .05$, reject H_0 at the $\alpha = .05$ significance level.

From these results, you have strong statistical evidence that the population proportion of all candy consumers who dies is less than the population proportion of all non-consumers who die.

b. You cannot conclude a cause-and-effect relationship between candy consumption and increased survival because this was an observational study and not a well-designed, randomized experiment.

c. You cannot generalize your conclusions to all adults because all the participants in this observational study were males. You should not generalize your conclusions to all males either because all the males in this study were men who attended an Ivy League school more than 30 years before the study. Therefore, they are not representative of the "typical" American male.

Exercise 21-29: Praising Children

a. The explanatory variable is *whether the child was praised for intelligence or for effort*.
The response variable is *whether the child lied about his or her score*.

b. This is an experiment because the researchers randomly determined whether the child would be praised for intelligence or for effort.

c. The sample proportions are $\hat{p}_{intelligence} = .379$ and $\hat{p}_{effort} = .133$.
These proportions do differ in the direction conjectured by the psychologists because the proportion of children praised for their intelligence who lied is almost three times as large as the proportion of children praised for their effort who lied.

d. To conduct a simulation for determining whether the difference in these conditional proportions is statistically significant, first mark 15 cards as "successes" (children who lied), and 44 cards as failures (children who did not lie). Then shuffle the cards well and deal 29 of the cards into an "intelligence" group, and put the remaining cards in an "effort" group. Calculate the difference in the proportion of successes in each group (intelligence group – effort group), and record this value. Repeat a large number of times. Then determine the total number of times the difference in the proportion of successes is at least .379 – .133 = .243 (the value seen in this study). Divide this total by the number of repetitions in order to estimate the *p*-value.

Note: You could focus on just the number of successes in the intelligence (or effort) group. In this case, in order to calculate the *p*-value you would count the number of times the number of successes in the intelligence group was at least 11, then divide by the total number of repetitions.

e. The *p*-value of .0298 measures the probability of finding at least 11 of the 15 liars in the "intelligence" group of 29 children by random assignment alone, assuming that praising children for intelligence and for effort are equally likely to result in a child lying about her or his score. In other words, if it makes no difference whether you praise a child for intelligence or effort, then you could expect to find at least 11 of the 15 liars among the 29 children praised for intelligence in only about 2.98% of the random assignments.

f. Because the *p*-value = .0298 < .05, these results are statistically significant at the α = .05 level.

g. You have moderate statistical evidence that praising a child for intelligence is more likely to result in the child lying about her or his score than praising a child for effort. Because this study was a well-designed, randomized experiment, you can conclude that praising for intelligence *causes* a child to lie about her or his score. You are not told how the children were selected for this study, so you should be cautious about generalizing to all children, or to children different from those in this sample.

Exercise 21-31: Nicotine Lozenge

a. The null hypothesis is that the population proportion of all smokers who could potentially use the nicotine lozenge who are male is the same as the population proportion of all smokers who could potentially use a placebo who are males. In symbols, H_0: $\pi_{\text{nicotine}} = \pi_{\text{placebo}}$
The alternative hypothesis is that the population proportion of all smokers who could potentially use the nicotine lozenge who are male is not the same as the population proportion of all smokers who could potentially use a placebo who are males. In symbols, H_a: $\pi_{\text{nicotine}} \neq \pi_{\text{placebo}}$.

Technical conditions: The values $n_1\hat{p}_c$, $n_1(1-\hat{p}_c)$, $n_2\hat{p}_c$, and $n_2(1-\hat{p}_c)$ are all greater than 10 (the smallest is 184), so this condition is met. The subjects were randomly assigned to two treatment groups, so this condition is also met.

The sample proportions of males in each group are $\hat{p}_{\text{nicotine}} = .429$ and $\hat{p}_{\text{placebo}} = .402$.

The test statistic is $z = \dfrac{.429 - .402}{\sqrt{(.415)(1-.415)\left(\dfrac{1}{459} + \dfrac{1}{458}\right)}} = 0.84$.

Using Table II, p-value $= 2 \times \Pr(Z > 0.84) = 2 \times .2005 = .4010$.

Because the p-value is not small, do not reject H_0 at any commonly used significant level.

You do not have sufficient statistical evidence to conclude there is a difference in the population proportion of males in the nicotine lozenge and placebo groups.

b. You suspect the researchers are pleased that this test produced a non-significant result because this means the observed difference in the proportions of each group who successfully abstained from smoking could not be attributed to the potentially confounding variable of gender.

Exercise 21-33: Native Californians

a. For a 95% CI, you calculate $(.516 - .438) \pm (1.96)\sqrt{\dfrac{(.516)(.484)}{500} + \dfrac{(.438)(.562)}{500}} = (.016, .140)$. You are 95% confident the 2000 percentage of Californians who were born in California is between 1.6 and 14 percentage points higher than the 1950 percentage of Californians who were born in California.

b. If you had subtracted in the other order, the interval would be $(-.140, -.016)$. The interpretation of this interval would not change.

Exercise 21-35: Pet CPR

a. For sample sizes $n_1 = n_2 = 100$, the test statistic is $z = \dfrac{.63 - .53}{\sqrt{(.58)(1-.58)\left(\dfrac{1}{100} + \dfrac{1}{100}\right)}} = 1.43$.

Using Table II, the p-value is $2 \times \Pr(Z > 1.43) = 2 \times .0764 = .1582$.

For sample sizes $n_1 = n_2 = 500$, the test statistic is $z = \dfrac{.63 - .53}{\sqrt{(.58)(1-.58)\left(\dfrac{1}{500} + \dfrac{1}{500}\right)}} = 3.20$.

Using Table II, the p-value is $2 \times \Pr(Z > 3.20) = 2 \times .0007 = .0014$.

For sample sizes $n_1 = 600$, $n_2 = 400$, the test statistic is $z = \dfrac{.63 - .53}{\sqrt{(.59)(1-.59)\left(\dfrac{1}{600} + \dfrac{1}{400}\right)}} = 3.14$.

Using Table II, the p-value is $2 \times \Pr(Z > 3.14) = 2 \times .0008 = .0016$.

Exercise 21-37: Hypothetical Medical Treatment Study

a. Here are the segmented bar graphs:

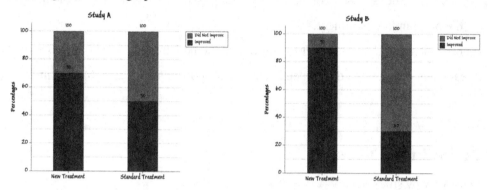

Study B shows a greater difference in improvement rate between the two treatments. The difference in improvement rate for study B is $.9 - .3 = .6$. The difference in improvement rate for study A is $.7 - .5 = .2$.

b. It would not be valid to use the two-sample z-test to compute the p-value for either study because the sample sizes in both studies are too small. (The values $n_1\hat{p}_c$, $n_1(1-\hat{p}_c)$, $n_2\hat{p}_c$, and $n_2(1-\hat{p}_c)$ are all less than 10.)

c. Results will vary. The following is from one representative running of the applet.

For study A, here are the applet results:

Two-way Table Inference

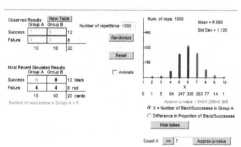

The approximate p-value is .345. This means you have not found convincing statistical evidence that the new treatment is any better than the old treatment.

For study B, here are the applet results:

Two-way Table Inference

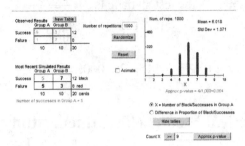

The approximate *p*-value is .004. You have very strong statistical evidence that the new treatment is better than the old standard treatment.

d. It makes sense that study B has stronger evidence that the new treatment is an improvement over the standard treatement because the difference in sample proportions (.9 – .3 = .6) is so much greater in study B than it is in study A (.7 – .5 = .2).

Exercise 21-39: In the News

Answers will vary by student.

Topic 22

Comparing Two Means

Odd- Numbered Exercise Solutions

Exercise 22-5: Properties of *p*-values

a. If all of the women sampled had one more close friend than they originally reported, the women's sample mean would increase by one, but the standard deviation would not change, and so the difference in the men's and women's sample means would also increase by one. This would increase the value of the test statistic and therefore decrease the *p*-value.

b. If all of the men sampled had one more close friend than they originally reported, the men's sample mean would increase by one, but their standard deviation would stay the same. In this case, the difference in the men's and women's sample means would *decrease* by one, and thus the test statistic would decrease. This means the *p*-value would increase.

c. If every man and every women sampled had one more close friend than they originally reported, the men's sample mean and the women's sample mean would both increase by one (but their standard deviations would not change). This means the difference in the men's and women's sample means would not change, so the *p*-value would not change either.

d. If both the men's and women's standard deviations were larger, the standard error, $\sqrt{\dfrac{s_1^2}{n_1} + \dfrac{s_2^2}{n_2}}$, would be larger, and therefore the test statistic would be smaller. A smaller test statistic would result in a larger *p*-value.

e. If both sample sizes were larger, the standard error would be smaller, and therefore the test statistic would increase. This would result in a smaller *p*-value.

Exercise 22-7: Hypothetical Commuting Times

Note that the differences in the sample medians are similar in each case but you don't have very clear information on how the means compare especially when there is skewness in the distributions. Manuel will have a large *p*-value because these sample sizes are small and the sample standard deviations are large. Notice the amount of overlap between the two sample distributions. Jacque's *p*-value will be less than Manuel's because the standard deviations are smaller. Katrina's *p*-value will be less than Manuel's because the sample sizes are larger (and the sample standard deviations are slightly smaller). (*Note*: It's not obvious how the *p*-values for Jacque and Katrina will compare.) Liam will have the smallest *p*-value because his sample sizes are the largest and his standard deviations are the smallest. Notice the small amount of overlap between the two sample distributions.

Exercise 22-9: Got a Tip?

a. The explanatory variable is *whether the party received a fancy piece of chocolate*.
    The response variable is the *amount of tip* (as a percentage of the bill).

b. Looking at the percentage of the bill is more useful than looking at the exact tip amount in this case because the checks would not all be for the same amount. Some parties would order more expensive meals and other might order very inexpensive meals. Thus, the amount of tip might vary considerably based on what the customers had for dinner, rather than on whether they had received the chocolate.

c. This is an experiment because the researchers randomly decided who would receive the chocolates (the treatment) and who would not (the control).

d. Technical conditions: Yes, you can check the technical conditions. The sample sizes are both greater than 30 ($n_1 = n_2 = 46$), and the data arise from random assignment of subjects to two treatment groups.

e. The null hypothesis is that providing a fancy, foil-wrapped piece of chocolate makes no difference in the average tip as a percentage of the bill at this restaurant (for all potential customers). In symbols, the null hypothesis is H_0: $\mu_{chocolate} = \mu_{no\ chocolate}$.

    The alternative hypothesis is that providing a fancy foil-wrapped piece of chocolate will increase the average tip as a percentage of the bill at this restaurant. In symbols, H_a: $\mu_{chocolate} > \mu_{no\ chocolate}$.

    The test statistic is $t = \dfrac{17.84 - 15.06}{\sqrt{\dfrac{(3.06)^2}{46} + \dfrac{(1.89)^2}{46}}} = 5.24$

    Using Table II with 40 degrees of freedom, you see that the *p*-value is off the chart, so *p*-value < .0005. Using Minitab and the applet, the *p*-value $\approx$ .0000.

    With such small *p*-value, reject H_0 at any common significance level. Because this was also a randomized experiment, you can conclude that there is strong statistical evidence that providing a fancy, foil-wrapped piece of chocolate will cause an increase the average tip (as a percentage of the bill) at this restaurant.

f. The confidence interval formula is $(17.84 - 15.06) \pm t^*(.5303)$. Using Table III with 40 degrees of freedom, $t^* = 2.021$, so the confidence interval is $(1.708, 3.852)$. Using Minitab, the interval is $(1.823, 3.837)$. Using the applet, the interval is $(1.712, 3.838)$. You are 95% confident that providing a fancy, foil-wrapped piece of chocolate will increase the average tip as a percentage of the bill by 1.7% to 3.8%.

g. Random assignment should have assured that the only difference between the parties was whether they received the chocolate with their bill, so you can conclude a causal link between the chocolate and the increased tip. You would not want to generalize this to other restaurants in other cities, or to other restaurants in Ithaca, as this experiment was tried at only one restaurant. It is possible that all 92 parties came to the restaurant on the same evening, in which case you should not generalize beyond that particular night of the week.

h. A Type I error would be concluding that the fancy, foil-wrapped piece of chocolate increases the average tip (as a percentage of the bill) at this restaurant, when in fact the chocolate makes no difference in the average tip. Such an error might mean that the manager and wait staff incur the expense of fancy chocolates without any return on their investment.

A Type II error would be failing to realize that a fancy, foil-wrapped piece of chocolate will increase the average tip as a percentage of the bill at this restaurant. Such an error would cost the wait staff this increase in tips as they would be unlikely to supply the fancy chocolates if they believe the chocolates have no effect on their tips.

Exercise 22-11: Ideal Age

a. The null hypothesis is that the mean age given by women is the same as the mean age given by men in response to this question. In symbols, $H_0: \mu_W = \mu_M$.

The alternative hypothesis is that the mean age given by women is not the same as the mean age given by men in response to this question. In symbols, $H_a: \mu_W \neq \mu_M$.

b. You need to know the standard deviations and how many in the sample were women and how many were men.

c. The test statistic is $t = \dfrac{43 - 39}{\sqrt{\dfrac{25^2}{1153} + \dfrac{25^2}{1153}}} = 3.84$.

Using Table II with 500 degrees of freedom, the *p*-value is off the chart, so *p*-value < .0005. Using Minitab, the *p*-value ≈ .0000. Using the applet, the *p*-value is .0001.

With such a small *p*-value, reject H_0 at any commonly used significance level. You have very strong statistical evidence that the mean age given in response to this question is not the same for women and men.

d. The confidence interval formula is $(43 - 39) \pm t^*(1.041)$. Using Table III with 500 degrees of freedom, $t^* = 2.586$, so the confidence interval is (1.308, 6.692). Using Minitab, the interval is (1.32, 6.68). Using the applet, the interval is (1.314, 6.686). You are 99% confident the average age given by women in response to this question is between 1.3 and 6.68 years greater than the average age given by men in response to this question.

e. The confidence interval is consistent with the test result because zero is not in the confidence interval, indicating that it is not a plausible value for the difference in average ages.

f. Answers will vary. Twenty-five is probably a reasonable upper bound for how high the standard deviation might be. It is hard to imagine that the "typical deviation from the mean" would be greater than 25 years for either gender.

Exercise 22-13: Editorial Styles

The *Washington Post* IQR is $28.5 - 16 = 12.5$, and $1.5 \times 12.5 = 18.75$, so $[Q_L - 18.75, Q_U + 18.75] = [0, 47.25]$. Therefore, remove any *Washington Post* sentence lengths outside this range. There are two such outliers (the sentences with lengths 56 and 50 words).

The following five-number summaries and boxplots compare the distributions of sentence length after these two observations have been removed:

	n	Min.	Q_L	Median	Q_U	Max	Mean	SD
USA Today	31	4	17	22	29	39	22.84	8.76
Washington Post	19	11	16	19	28.5	41	20.79	8.29

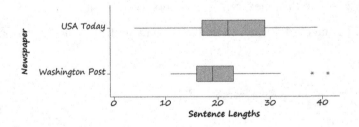

The null hypothesis is that the average length of (all) sentences in both papers is the same. In symbols, H_0: $\mu_{\text{USA Today}} = \mu_{\text{W Post}}$.

The alternative hypothesis is that average length of all sentences in *USA Today* is less than the average length of all sentences in the *Washington Post*. In symbols, H_a: $\mu_{\text{USA Today}} < \mu_{\text{W Post}}$.

The test statistic is $t = \dfrac{22.84 - 20.79}{\sqrt{\dfrac{(8.76)^2}{31} + \dfrac{(8.29)^2}{19}}} = 0.83$.

Using Table III with 18 degrees of freedom, $-0.862 < -0.32$, so the *p*-value > .20. Using Minitab, the *p*-value is .794. Using the applet, the *p*-value is .7914.

Test decision: With the large *p*-value, do not reject H_0 at any commonly used significance level.

Conclusion in context: You do not have any statistical evidence to suggest that the average length of (all) the sentences in *USA Today* is less than the average length of the sentences in the *Washington Post*.

This time there is no possibility of rejecting the null hypothesis, because without the outliers, the sample mean sentence length for the *Washington Post* is *less* than the sample mean sentence length for *USA Today*. So the sample provides no evidence that *USA Today* has shorter sentences than the *Washington Post*.

Exercise 22-15: Children's Television Viewing

a. *Baseline Videotapes*:

H_0: $\mu_{\text{control}} = \mu_{\text{intervention}}$ vs. H_a: $\mu_{\text{control}} \neq \mu_{\text{intervention}}$

The test statistic is $t = \dfrac{5.52 - 4.74}{\sqrt{\dfrac{(10.44)^2}{103} + \dfrac{(6.57)^2}{95}}} = 0.63$.

Using Table III with 80 degrees of freedom, $0.63 < 0.846$, so *p*-value $> 2 \times .20 = .40$. Using Minitab, the *p*-value is .527. Using the applet, the *p*-value is .5275.

With this large *p*-value, do not reject H_0 at any common significance level.

You do not have statistically significant evidence of a difference in the baseline videotape group population means.

Baseline Video Games:

H_0: $\mu_{\text{control}} = \mu_{\text{intervention}}$ vs. H_a: $\mu_{\text{control}} \neq \mu_{\text{intervention}}$

The test statistic is $t = \dfrac{3.85 - 2.57}{\sqrt{\dfrac{(9.17)^2}{103} + \dfrac{(5.1)^2}{95}}} = 1.23$.

Using Table III with 80 degrees of freedom, $0.846 < 1.23 < 1.292$, so $2 \times .10 < p\text{-value} < 2 \times .20$, which means $.20 < p\text{-value} < .40$ Using Minitab, the p-value is .222. Using the applet, the p-value is .2233.

With this large p-value, do not reject H_0 at any common significance level.

You do not have statistically significant evidence of a difference in the baseline video game group population means.

b. *Follow up Videotapes*:

$H_0: \mu_{control} = \mu_{intervention}$ vs. $H_a: \mu_{control} > \mu_{intervention}$

The test statistic is $t = \dfrac{5.21 - 3.46}{\sqrt{\dfrac{(98.41)^2}{103} + \dfrac{(4.86)^2}{95}}} = 1.81$.

Using Table III with 80 degrees of freedom, $1.664 < 1.81 < 1.990$, so $.025 < p\text{-value} < .05$. Using Minitab, the p-value is .036. Using the applet, the p-value is .0368.

Because the p-value is less than .05, reject H_0 at the $\alpha = .05$ significance level.

You have moderate statistical evidence that the population mean time spent watching videotapes is greater under the control condition than under the intervention condition.

Follow up Video Games:

$H_0: \mu_{control} = \mu_{intervention}$ vs. $H_a: \mu_{control} > \mu_{intervention}$

The test statistic is $t = \dfrac{4.24 - 1.32}{\sqrt{\dfrac{(10)^2}{103} + \dfrac{(2.72)^2}{95}}} = 2.85$.

Using Table III with 80 degrees of freedom, $2.639 < 2.85 < 3.195$, so $.001 < p\text{-value} < .005$. Using Minitab, the p-value is .003. Using the applet, the p-value is .0027.

With this small p-value, reject H_0 at any commonly used significance level.

You have strong statistical evidence that the population mean time spent playing video games is greater under the control condition than under the intervention condition.

Exercise 22-17: UFO Sighters' Personalities

a. This is an observational study. The researchers did not determine who would or would not be UFO sighters.

b. The explanatory variable is *whether the subject had an intense experience with a UFO*. This variable is binary categorical. The response variable is *IQ score*. This variable is quantitative.

c. The null hypothesis is that population mean IQ score is the same for all community members and all UFO sighters. In symbols, H_0: $\mu_{community} = \mu_{UFO}$.

The alternative hypothesis is that the population mean IQ score for all community members is not the

Same as the population mean IQ for all UFO sighters. In symbols, H_a: $\mu_{community} \neq \mu_{UFO}$.

The test statistic is $t = \dfrac{101.6 - 100.6}{\sqrt{\dfrac{(8.9)^2}{25} + \dfrac{(12.3)^2}{53}}} = 0.41$.

Using Table III with 24 degrees of freedom, $0.41 < 0.857$, so the *p*-value $> 2 \times .2 = .4$.

Using Minitab, the *p*-value is .684. Using the applet, the *p*-value is .6873.

The *p*-value is not small, so do not reject H_0. There is not sufficient statistical evidence to suggest the population mean IQ scores of community members differs from that of UFO sighters.

d. The 95% confidence interval for $\mu_{UFO} - \mu_{community}$ is $(101.6 - 100.6) \pm t^*(2.45)$.

Using Table III with 24 degrees of freedom, t^* - 2.064, so the confidence interval is $(-4.057, 6.057)$.

Using Minitab, the interval is $(-3.9, 5.9)$. Using the applet, the interval is $(-4.065, 6.065)$.

You are 95% confident the difference in the mean IQ score between these two populations is between -4 and 6 points. Because this interval contains both positive and negative numbers and zero, it indicates that it is plausible there is no difference in the mean IQ scores of the two populations. Each of the values from -4 to 6 (including zero) is a plausible value for the difference in mean IQ scores between these two populations.

e. Even if you had found a significantly higher mean IQ for one group, you would not be able to draw any causal conclusions about how seeing UFOs affects intelligence because this was an observational study. There would be many potential confounding variables for which you were unable to control.

Exercise 22-19: Ice Cream Servings

a. This is an experiment because the researchers randomly assigned the subjects to two bowl sizes.

b. Technical conditions: No, you do not have enough information to check whether the technical conditions of the two-sample t-test are satisfied. The sample sizes are small ($n_1 = 17 < n_2 = 20 < 30$), and you do not have the individual data, so you cannot make a judgment about whether the data appear to come from normally distributed populations.

c. The null hypothesis is that population mean volume of ice cream taken by people is the same regardless of bowl size. In symbols, H_0: $\mu_{34 \text{ oz bowl}} = \mu_{17 \text{ oz bowl}}$.

The alternative hypothesis is that the population mean volume of ice cream taken by people with a 34-ounce bowl is greater than the mean volume of ice cream taken by people with a 17-ounce bowl. In symbols, H_a: $\mu_{34 \text{ oz bowl}} > \mu_{17 \text{ oz bowl}}$.

d. A Type I error would be concluding that a 34-ounce bowl causes people to take more ice cream, on average, than people who have a 17-ounce bowl, when the size of the bowl actually has no effect on the population mean amount of ice cream taken.

A Type II error would be failing to realize that a larger bowl size causes people to take more ice cream, on average, than a smaller bowl, when in fact the larger bowl does have this effect.

e. The test statistic is $t = \dfrac{5.81 - 4.38}{\sqrt{\dfrac{(2.26)^2}{17} + \dfrac{(2.05)^2}{20}}} = 2.00$

Using Table III with 16 degrees of freedom, $1.746 < 2.00 < 2.120$, so $.025 < p\text{-value} < .05$. Using Minitab, the p-value is .027. Using the applet, the p-value is .0313.

f. The difference in volume of ice cream served between the two bowl sizes is statistically significant at the .05 level, but not at the .01 level ($.01 < p\text{-value} < .05$).

g. You can draw a cause-and-effect conclusion between bowl size and volume of ice cream taken because this is a well-designed experiment and is significant at the 5% level. If the randomization worked, the only difference between the two groups should have been the bowl sizes and the small p-value indicates the randomization process alone is not a plausible explanation for the observed difference in the samples means.

Exercise 22-21: Natural Selection

a. The null hypothesis is that population mean total length of the sparrows that would survive such a storm is the same as the mean total length of the sparrows that would perish. In symbols, H_0: $\mu_{\text{die}} = \mu_{\text{survive}}$.

The alternative hypothesis is that population mean total length of the sparrows that would survive such a storm is not the same as the mean total length of the sparrows that would perish. In symbols, H_a: $\mu_{die} \neq \mu_{survive}$.

Technical conditions: Both sample sizes are not greater than 30 ($n_1 = 24$), but probability plots indicate it is plausible that both populations are normally distributed. However, the data do not appear to be independent random samples from two populations. You might be willing to consider them representative enough to proceed with the analysis.

The test statistic is $t = \dfrac{162 - 159.06}{\sqrt{\dfrac{(2.41)^2}{24} + \dfrac{(2.81)^2}{35}}} = 4.3$.

Using Table III with 23 degrees of freedom, $3.768 < 4.3$, so p-value $< 2 \times .0005$, which means p-value $< .001$. Using Minitab and the applet, p-value $\approx .0000$.

With such a small p-value, reject H_0 at any commonly used significance level and conclude there is a difference in the population mean total lengths between sparrows that would die and the sparrows that would survive a severe winter storm (as long as the sparrows are representative of these populations). A 95% confidence interval for $\mu_{die} - \mu_{survive}$ is $(162 - 159.06) \pm t^*(.684)$.

Using Table II with 23 degrees of freedom, $t^* = 2.069$ so the confidence interval is $(1.525, 4.355)$. Using Minitab, the interval is $(1.572, 4.314)$. Using the applet, the interval is $(1.525, 4.355)$.

You are 95% confident the population average length of the sparrows that would die in such a storm is between 1.5 and 4.3 mm greater than the population average length of the sparrows that would survive.

Because this was an observational study and not an experiment, you cannot conclude a cause-and – effect relationship between total length and the death of the sparrows. In addition, because these data were not randomly selected and you cannot tell whether they are representative, you probably should not generalize these results beyond sparrows in Providence, Rhode Island, in the late 1800s.

b. The null hypothesis is that population mean weight of the sparrows that would survive such a storm is the same as the mean weight of the sparrows that would perish. In symbols, H_0: $\mu_{die} = \mu_{survive}$.

The alternative hypothesis is that population mean weight of the sparrows that would survive such a storm is not the same as the mean weight of the sparrows that would perish.

In symbols, H_a: $\mu_{die} \neq \mu_{survive}$.

The test statistic is $t = \dfrac{26.27 - 25.47}{\sqrt{\dfrac{(1.46)^2}{24} + \dfrac{(1.26)^2}{35}}} = 2.19$.

Using Table III with 23 degrees of freedom, $2.069 < 2.19 < 2.500$, so $2 \times .01 < p\text{-value} < 2 \times .025$, which means $.02 < p\text{-value} < .05$. Using Minitab, the p-value is .034. Using the applet, the p-value is .0394.

With a p-value less than .05, reject H_0 at the $\alpha = .05$ significance level and conclude there is a difference in the population mean weight between sparrows that would die and the sparrows that would survive a severe winter storm.

A 95% confidence interval for $\mu_{die} - \mu_{survive}$ is $(26.27\text{-}25.47) \pm t^*(.366)$.

Using Table II with 23 degrees of freedom, $t^* = 2.069$ so the confidence interval is $(.0427, 1.557)$. Using Minitab, the interval is $(0.064, 1.541)$. Using the applet, the interval is $(0.042, 1.558)$.

You are 95% confident the population mean weight of the sparrows that would die in such a storm is between .06 grams and 1.5 grams greater than the population mean weight of the sparrows that would survive.

c. *Alar extent*

H_0: $\mu_{die} = \mu_{survive}$ vs. H_a: $\mu_{die} \neq \mu_{survive}$

The test statistic is $t = \dfrac{247 - 247.69}{\sqrt{\dfrac{(3.79)^2}{24} + \dfrac{(3.86)^2}{35}}} = -0.68$.

Using Table III with 23 degrees of freedom, $-0.858 < -0.68$, the p-value is off the chart. This means the p-value $> 2 \times .20 = .40$. Using Minitab, the p-value is .501. Using the applet, the p-value is .5022. Because the p-value is not small, do not reject H_0. There is not convincing evidence of a difference in the mean alar extent between sparrows that would die and the sparrows that would survive a severe winter storm.

A 95% confidence interval for $\mu_{die} - \mu_{survive}$ is $(247 - 247.69) \pm t^*(1.01)$.

Using Table II with 23 degrees of freedom, $t^* = 2.069$ so the confidence interval is $(-2.784. 1.404)$. Using Minitab, the interval is $(-2.72, 1.35)$. Using the applet, the interval is $(-2.784. 1.404)$.

You are 95% confident the population mean alar extent is between 2.784 less to 1.404 greater for the sparrows that would die in such a storm than for the sparrows that would survive such a storm.

Length of beak and head

H_0: $\mu_{die} = \mu_{survive}$ vs. H_a: $\mu_{die} \neq \mu_{survive}$

The test statistic is $t = \dfrac{31.671 - 31.614}{\sqrt{\dfrac{(.611)^2}{24} + \dfrac{(.631)^2}{35}}} = 0.34.$

Using Table III with 23 degrees of freedom, $0.34 < 0.858$, so the *p*-value is off the chart. This means the *p*-value $> 2 \times .20 = .40$. Using Minitab, the *p*-value is .732. Using the applet, the *p*-value is .7315. Because the *p*-value is not small, do not reject H_0.

There is not convincing evidence of a difference in the mean length of beak and head between sparrows that would die and the sparrows that would survive a severe winter storm.

A 95% confidence interval for $\mu_{die} - \mu_{survive}$ is $(31.671 - 31.614) \pm t^*(1.64)$.

Using Table II with 23 degrees of freedom, $t^* = 2.069$ so the confidence interval is $(-0.283, 0.397)$. Using Minitab, the interval is $(-0.273, 0.386)$. Using the applet, the interval is $(-0.282\ 0.396)$.

You are 95% confident the population mean length of beak and head is between .282 less to .397 greater for the sparrows that would die in such a storm than for the sparrows that would survive such a storm.

Humerus bone length

H_0: $\mu_{die} = \mu_{survive}$ vs. H_a: $\mu_{die} \neq \mu_{survive}$

The test statistic is $t = \dfrac{.7279 - .738}{\sqrt{\dfrac{(.0235)^2}{24} + \dfrac{(.0198)^2}{35}}} = -1.72.$

Using Table III with 23 degrees of freedom, $-2.069 < -1.72 < -1.714$, so $2 \times .025 < p\text{-value} < 2 \times .05$, which means $.05 < p\text{-value} < .10$. Using Minitab, the *p*-value is .092. Using the applet, the *p*-value is .0976.

Because the *p*-value is greater than .05, do not reject H_0 at the 5% level (but you would reject at the 10% level). There is not enough statistical evidence at the 5% level to conclude there is a difference in the mean humerus bone length between sparrows that would die and the sparrows that would survive a severe winter storm.

A 95% confidence interval for $\mu_{die} - \mu_{survive}$ is $(.7269 - .738) \pm t^*(.0058)$.

Using Table II with 23 degrees of freedom, $t^* = 2.069$ so the confidence interval is $(-0.022, 0.002)$. Using Minitab, the interval is $(-0.0219, 0.00173)$. Using the applet, the interval is $(-0.022\ 0.002)$.

You are 95% confident the population mean humerus bone length is between .022 less to .022 greater for the sparrows that would die in such a storm than for the sparrows that would survive such a storm.

Femur bone length

$H_0: \mu_{die} = \mu_{survive}$ vs. $H_a: \mu_{die} \neq \mu_{survive}$

The test statistic is $t = \dfrac{.7065 - .7168}{\sqrt{\dfrac{(.0203)^2}{24} + \dfrac{(.0225)^2}{35}}} = -1.84$.

Using Table III with 23 degrees of freedom, $-2.069 < -1.84 < -1.714$, so $2 \times .025 < p\text{-value} < 2 \times .05$, which means the $.05 < p\text{-value} < .10$. Using Minitab, the p-value is .092. Using the applet, the

p-value is .0800.

Because the p-value is greater than .05, do not reject H_0 at the 5% level (but you would reject at the 10% level).

There is not enough statistical evidence at the 5% level to conclude there is a difference in the mean femur bone length between sparrows that would die and the sparrows that would survive a severe winter storm.

A 95% confidence interval for $\mu_{die} - \mu_{survive}$ is $(31.671 - 31.614) \pm t^*(1.64)$.

Using Table II with 23 degrees of freedom, $t^* = 2.069$ so the confidence interval is $(-0.283, 0.397)$. Using Minitab, the interval is $(-0.273, 0.386)$. Using the applet, the interval is $(-0.282, 0.396)$.

You are 95% confident the population mean femur bone length is between 2.82 less .396 greater for the sparrows that would die in such a storm than for the sparrows that would survive such a storm.

Tibiotarsus bone length

$H_0: \mu_{die} = \mu_{survive}$ vs. $H_a: \mu_{die} \neq \mu_{survive}$

The test statistic is $t = \dfrac{1.1202 - 1.1353}{\sqrt{\dfrac{(.0377)^2}{24} + \dfrac{(.036)^2}{35}}} = -1.55$.

Using Table III with 23 degrees of freedom, $-1.714 < -1.55 < -2.069$, so $2 \times .025 < p\text{-value} < 2 \times .05$, which means $.05 < p\text{-value} < .10$. Using Minitab, the p-value is .129. Using the applet, the p-value is .1325.

Because the p-value is greater than .05, do not reject H_0 at the 5% level. There is not enough statistical evidence to conclude there is a difference in the mean tibiotarasus bone length between sparrows that would die and sparrows that would survive a severe winter storm.

A 95% confidence interval for $\mu_{die} - \mu_{survive}$ e is $(1.1202 - 1.1353) \pm t^*(.0098)$.

Using Table II with 23 degrees of freedom, $t^* = 2.069$ so the confidence interval is $(-0.035, 0.005)$. Using Minitab, the interval is $(-0.03492, 0.00457)$. Using the applet, the interval is $(-0.036\ 0.005)$. You are 95% confident the population mean tibiotarsus bone length is between .036 less to .005 greater for the sparrows that would die in such a storm than for the sparrows that would survive such a storm.

Skull width

H_0: $\mu_{die} = \mu_{survive}$ vs. H_a: $\mu_{die} \neq \mu_{survive}$

The test statistic is $t = \dfrac{.6036 - .6025}{\sqrt{\dfrac{(.0126)^2}{24} + \dfrac{(.0139)^2}{35}}} = 0.33$.

Using Table III with 23 degrees of freedom, $0.33 < 0.858$, the p-value is off the chart. This means the p-value $> 2 \times .20 = .40$. Using Minitab, the p-value is .745. Using the applet, the p-value is .755. Because the p-value is not small, do not reject H_0.

There is no evidence of a difference in the mean skull width between sparrows that would die and the sparrows that would survive a severe winter storm.

A 95% confidence interval for $\mu_{die} - \mu_{survive}$ is $(.6036 - .6025) \pm t^*(.00343)$.

Using Table II with 23 degrees of freedom, $t^* = 2.069$ so the confidence interval is $(-0.006, 0.008)$. Using Minitab, the interval is $(-0.00585, 0.00813)$. Using the applet, the interval is $(-0.006\ 0.008)$. You are 95% confident the population skull width is between .006 less to .008 greater for the sparrows that would die in such a storm than for the sparrows that would survive such a storm.

Keel of sternum

H_0: $\mu_{die} = \mu_{survive}$ vs. H_a: $\mu_{die} \neq \mu_{survive}$

The test statistic is $t = \dfrac{.8457 - .8576}{\sqrt{\dfrac{(.0332)^2}{24} + \dfrac{(.0372)^2}{35}}} = -1.28$.

Using Table III with 23 degrees of freedom, $-1.319 < -1.28 < -0.858$, $2 \times .1 < p$-value $< 2 \times .2$, which means $.02 < p$-value $< .4$. Using Minitab, the p-value is .207. Using the applet, the p-value is

.2108.

Because the *p*-value is not small, do not reject H_0.

There is not convincing evidence of a difference in the mean keel of sternum between sparrows that would die and the sparrows that would survive a severe winter storm.

A 95% confidence interval for $\mu_{die} - \mu_{survive}$ is $(.8457 - .8576) \pm t^*(1.64)$.

Using Table II with 23 degrees of freedom, $t^* = 2.069$ so the confidence interval is $(-0.031, 0.007)$.

Using Minitab, the interval is $(0.0304, 0.0067)$. Using the applet, the interval is $(-0.031\ 0.007)$.

You are 95% confident the population mean keel of sternum is between .031 less to .007 greater for the sparrows that would die in such a storm than for the sparrows that would survive such a storm.

Exercise 22-23: Hypothetical SAT Coaching

a. The null hypothesis is that the population mean improvement in SAT scores would be the same for both the coaching and control groups. In symbols, H_0: $\mu_{coaching} = \mu_{control}$.

The alternative hypothesis is that the population mean improvement in SAT scores would be greater for those who are coached than for those who are not coached. In symbols, H_a: $\mu_{coaching} > \mu_{control}$.

The test statistic is $t = \dfrac{46.2 - 44.4}{\sqrt{\dfrac{(14.4)^2}{2500} + \dfrac{(15.3)^2}{2500}}} = 4.28$

Using Table III with 500 degrees of freedom, $4.28 > 3.310$, so *p*-value $< .0005$. Using Minitab and the applet, *p*-value $\approx .0000$.

With such a small *p*-value, reject H_0 at any commonly used significance level.

You have extremely strong statistical evidence that the population mean improvement in SAT scores when coached is greater than the population mean improvement in SAT scores when not coached.

b. For a 99% CI, you calculate $(46.2 - 44.4) \pm (2.576)(.420) = (.717, 2.883)$. You are 99% confident the population mean improvement in SAT scores when coached is between .717 and 2.883 points greater than the population mean improvement when not coached.

c. The data do provide very strong evidence that SAT coaching is helpful. The *p*-value is what helps you answer this question. The *p*-value is very small (≈ 0), indicating that you would essentially never see a sample result like this by chance alone if the SAT coaching had no effect.

d. No, the data indicate that SAT coaching will only improve SAT scores by at most 2.9 points on average, which means the coaching is not very helpful. You can tell this from the confidence interval.

Exercise 22-25: Hypothetical ATM Withdrawals

a. For a 90% CI, you calculate $(70 - 70) \pm t^*(6.06) = (-\$10.6, \$10.6)$.

b. This interval does include the value zero.

c. No, this interval would not be any different if you had chosen a different pair of machines because the sample means, sample sizes, and sample standard deviations are the same for all three machines.

d. No, in spite of the common sample sizes, means and standard deviations, as you learned in Exercises 9-24 and 19-21, these three machines are not identical in their distributions of withdrawal amounts. (See the description in Exercise 19-21, part a.) The two-sample t-test does not allow you to say anything about whether the *shapes* of the distributions differ – only that the means do or do not differ.

Exercise 22-27: Interpreting Statistical Significance

a. A small p-value (in a two-sample t-test) does not necessarily mean that all the data values in one sample are greater than all of the data values in the other sample. The small p-value indicates that the *mean* of one sample is significantly greater than the *mean* of the other.

b. A small p-value does not say anything about whether a cause-and-effect relationship exists between the explanatory and response variables. Such a relationship can only be determined by a well-designed, comparative, randomized experiment (in conjunction with a small p-value). A small p-value found from an observational study only tells you there is an association between the explanatory and response variables; there may be confounding variables that explain the association.

c. A small p-value does not indicate that the data resulted from either random sampling or random assignment. A test of significance is only valid if the data resulted from one of these, but the test statistic and p-value can be computed no matter how the data were collected.

d. A small p-value does not necessarily mean there were no outliers in the data. If the sample sizes are large, the population distributions do not need to be normally distributed (and the p-value can be calculated regardless of whether this technical condition is met). If the sample sizes are large, even samples with large standard deviations caused by outliers can result in a statistically significant difference in the population means.

e. The p-value only tells you that there is a difference in the population means. It does not tell you how large the difference is, or whether the difference is significant in any practical sense. You need a confidence interval in order to estimate the size of the difference and subject matter knowledge in order to decide whether it is an interesting difference.

Exercise 22-29: Influencing Charitable Donations

a. The observational units are the 118 college students who were given $5 for filling out a survey. The explanatory variable is *whether the student was shown the little girl or the statistical information.* The response variable was *whether the student donated some of the $5.*

b. This study is an experiment because the researchers randomly determined which students would be shown the little girl and which would be given the statistical information.

c. Both of these distributions are likely to be skewed to the right because the sample standard deviations are roughly equal to, or larger than the sample means. If you consider values in an interval one standard deviation below the mean, some of these values would be negative (in both distributions), but that makes no sense in this context. It is impossible to donate a negative amount of money. So these distributions must be skewed to the right.

d. The technical conditions for a 2-sample *t*-test are satisfied because both sample sizes are large ($n_1 = n_2$ = 59 > 30), and all subjects were randomly assigned to the two treatment groups.

e. The null hypothesis is that the population mean amount that would be donated in response to specific information about a particular victim is the same as the population mean amount that would be donated in response to statistical information about the extent of the larger problem. In symbols, H_0:

$\mu_{\text{little girl}} = \mu_{\text{statistical information}}$.

The alternative hypothesis is that the population mean amount that would be donated in response to specific information about a particular victim is greater than the population mean amount that would be donated in response to statistical information about the extent of the larger problem. In symbols,

H_a: $\mu_{\text{little girl}} > \mu_{\text{statistical information}}$.

The test statistic is $t = \dfrac{2.12 - 1.21}{\sqrt{\dfrac{(2.13)^2}{59} + \dfrac{(1.67)^2}{59}}} = 2.58$.

Using Table III with 50 degrees of freedom, $2.403 < 2.58 < 2.678$, so $2 \times .005 < p\text{-value} < 2 \times .01$, which means $.01 < p\text{-value} < .02$. Using Minitab, the *p*-value is .011. Using the applet, the *p*-value is .0062.

Because this *p*-value is less than .05, reject H_0 at the 5% significance level.

c. For a 99% CI, you calculate $(2.12 - 1.21) \pm t^*(0.352)$. Using Table III with 27 degrees of freedom, t^* = 2.678, so the confidence interval is (–0.034, 1.854). Using Minitab, the interval is (–0.014, 1.834). Using the applet, the interval is (–0.028, 1.848).

You are 99% confident the population mean amount that would be donated in response to specific information about a particular victim is between –$.03 and $1.85 more than the population mean amount that would be donated in response to statistical information about the extent of the larger problem.

d. You have moderate statistical evidence that the population amount that would be donated in response to specific information about a particular victim is between $.03 less to $1.85 more, with 99% confidence, than the population mean amount that would be donated in response to statistical information about the extent of the larger problem. Because this was a well-designed experiment, you can conclude that viewing specific information about a particular victim causes the increase in donations. However, because you do not know how these college students were selected, you should be cautious in generalizing these results beyond students from these colleges.

Exercise 22-31: M&M Consumption

a. A 90% confidence interval for $\mu_F - \mu_M$ is $(56.7 - 43.6) \pm t^*(13.58)$.

Using Table III with 9 degrees of freedom, $t^* = 1.833$, so the confidence interval is $(-11.79, 37.99)$. Using Minitab, the interval is $(-10.5, 36.7)$, and using the applet, the interval is $(-11.794, 37.994)$.

b. You are 90% confident the mean number of M&Ms taken by females from such a bowl is between –11.8 and 38 more than the mean number of M&Ms taken by males.

c. Yes, the confidence interval and test decision from Exercise 22-30 are consistent. The confidence interval contains both positive and negative numbers as well as zero, which tells you that zero is a plausible value for the difference in the mean number of M&Ms taken by females and males. The test decision is that there is not sufficient statistical evidence of a difference between the number of M&Ms taken by males and females, which means that difference in the number of M&Ms taken could plausibly be zero.

Exercise 22-33: Culture and Academics

a. The explanatory variable in this study is *whether the student was asked questions about gender or race*.

The response variable is *performance on the math test*.

b. The population of interest is all female Asian-American college students.

c. This study made use of random assignment; the students were randomly assigned to the "gender questions" and "race questions" groups.

d. The null hypothesis is that the population mean score on the math test is the same for all students in the "gender questions" treatment as in the "race questions" treatment. In symbols, H_0: $\mu_{Gender} = \mu_{Race}$. The null hypothesis is that the population mean score on the math test is greater for the questions asked about race treatment than for the questions asked about gender treatment. In symbols, H_0: μ_{Race} > μ_{Gender}.

e. The fact that this study was reported to be statistically significant means that the probability of obtaining sample results as or more extreme as these, if the questions asked had no effect on the math test performance, was small. It would be unlikely to see these sample results if the population mean score on the math test was the same for both the "gender questions" treatment and the "race questions" treatment.

f. A cause-and-effect conclusion *is* justified in this study because the researcher performed a well-designed, randomized experiment by randomly assigning the subject to the two question groups.

g. Because this study was a well-designed, randomized experiment, and the results were statistically significant, you can conclude that asking students questions about race causes them to perform better on this math test than asking students questions about gender does. In addition, you can conclude that asking students about gender causes them to perform worse on the math test than asking students about race.

Exercise 22-35: Sleep Deprivation

Answers will vary by student. The following are based on one representative running of the applet. The difference in the sample group medians is –12.05. The applet was used to perform 1000 randomizations of the 21 improvement scores into the two groups, calculating the difference in sample medians each time. The resulting dotplot of the differences in sample medians was roughly symmetric (normal) with a center (mean) of –0.21, and a standard deviation of 5.65. To estimate the *p*-value, the applet computed the number of the differences (in medians) that were ≤ -12.05. This provided an approximate *p*-value of $27/1000 = .0270$.

Based on this approximate *p*-value, the results are statistically significant at the $\alpha = .05$ level. If there were no difference in the population medians, you would obtain a difference in sample medians of –12.05 or less in roughly 2.7% of all random assignments. Because this difference in medians actually did occur in the study, you conclude that there is convincing evidence subjects in the sleep deprivation treatment have a lower population median improvement time than do subjects in the unrestricted sleep treatment.

Exercise 22-37: M&M Consumption

a. Results will vary. The following are from one representative running of the applet with 1000 repetitions.

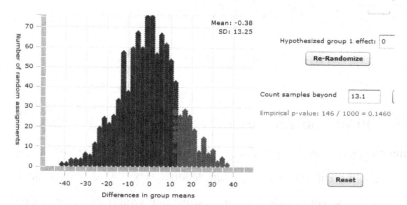

The applet reports an approximate *p*-value of 146/1000 = .1460. Because this *p*-value is large, do not reject H_0 at any commonly used significance level. You have not found convincing statistical evidence that the population mean number of M&Ms taken by females from such a bowl differs from the population mean number of M&Ms taken by males.

b. This *p*-value is approximately half of the *p*-value found in Exercise 22-30, although the test decisions are the same. This makes sense because the Randomization applet is performing a simulation for a one-sided test of H_0: $\mu_F = \mu_M$ vs. H_a: $\mu_F > \mu_M$ whereas the test in Exercise 22-30 was two-sided.

Exercise 22-39: Own Your Own

Answers will vary by student.

Topic 23

Analyzing Paired Data

Odd- Numbered Exercise Solutions

Exercise 23-5: Culture and Academics

a. To create a matched pairs design to address this research question, you would have to find some traits from which you could create pairs of one Chinese student one American student. This would be useful if you thought you could pair them up in ways that would make them similar to each other apart from their nationality. However it might be difficult to know what traits would be useful in this context. For example, though you could pair them by height, that is probably not useful in reducing variability in counting ability among 4-year-olds.

b. You could address this research question with a matched-pairs design by having each student take the math test twice – once after being asked the questions about race, and once after being asked the questions about gender. You would need to randomize the order in which the two questions were asked, and you would need to allow for the fact that all the students are likely to perform better on the math test the second time they see it.

Exercise 23-7: Car Ages

a. This study calls for an independent-samples analysis. There is no link between the students and the faculty members in the sample. In this particular case, you could not do matched-pairs analysis because the sample sizes are different.

b. No, the answer would not change if you sampled 25 students and 25 faculty members. There would still be no link between the students and faculty members, they would still be *independent* samples.

Exercise 23-9: Sickle Cell Anemia and Child Development

a. This is an observational study. The researchers did not determine which children would/would not have sickle cell anemia.

b. This is a matched-pairs study. Each child with the sickle cell trait was matched with a child in the control group.

c. The *p*-values must be large (greater than .10) if this study found no statistically significant differences. The researchers would not have rejected any of their null hypotheses in this study.

Exercise 23-11: Friendly Observers

Have each subject play the video game twice – once when they know an observer with a vested interest is watching and once when they know an observer with no vested interest is watching. Be sure to randomly decide the order in which the type of observer is assigned to each playing of the game. Record the difference in the times it takes to navigate the obstacle course in the video game.

The primary advantage to redoing this experiment with a matched-pairs design is that you could reduce the overall variability (eliminate the subject-to-subject variability to focus on the effect of the observers' interest level).

Exercise 23-13: Marriage Ages

a. It would be appropriate to use a matched-pairs analysis for the first way of collecting data (take a random sample of couples and record their ages) because you already have pairs of husbands and wives matched. It would *not* be appropriate to use a matched-pairs analysis for the second way of collecting the data (take a random sample of wives and a separate random sample of husbands) because these wives and husbands are not matched with each other.

b. It is not possible to conduct a randomized experiment on this issue because you (the researcher) cannot assign the explanatory variable (husband or wife) to the subjects in the study. That variable is pre-determined.

c. Taking a large random sample of couples would be preferable to taking a small random sample of couples because a larger sample size would result in a smaller standard error ($s/\sqrt{n}$). This would make any confidence interval more narrow, and therefore make for a more precise estimate of the population mean difference in married couples' ages. A larger sample size would also increase the power of any significance test, which makes it more likely that you would detect an existing mean difference in couples' ages.

Exercise 23-15: Catnip Aggression

a. The null hypothesis is that the population mean number of negative cat interactions is the same before and after exposure to catnip. In symbols, H_0: $\mu_{before} = \mu_{after}$.

The alternative hypothesis is that the population mean number of negative cat interactions is not the same before exposure to catnip as it is after. In symbols, H_a: $\mu_{before} \neq \mu_{after}$.

The test statistic is $t = \dfrac{1.8 - 2.8}{\sqrt{\dfrac{(1.66)^2}{15} + \dfrac{(2.37)^2}{15}}} = -1.34$.

Using Table III with 14 degrees of freedom, $-1.345 < -1.34 < -0.868$, so $2 \times .1 < p$-value $< 2 \times .2$. This means $.20 < p$-value $< .40$. Using Minitab, the p-value is .192. Using the applet, the p-value is .2021.

With a p-value of approximately .2, do not reject H_0 at any common significance level.

You do not have sufficient statistical evidence to conclude there is a difference in the mean number of negative cat interactions before and after exposure to catnip.

b. The p-value has increased to the point where you cannot reject H_0 at any reasonable significance level.

c. Yes, this analysis indicates the pairing was useful. The variability was significantly reduced by using paired data, to account for the cat-to-cat variation in the amount of negative cat interactions.

Exercise 23-17: Mice Cooling

a. These data call for a matched-pairs analysis because each mouse's body was used twice to measure the cooling constant; once when freshly killed and once when reheated.

b. Here is the numerical summary and graphical display for the differences (*fresh − reheated*):

	n	$\bar{x}$	s	Min.	Q_L	Median	Q_U	Max.
Differences	19	20.8	45.8	−45	−11	10	45	139

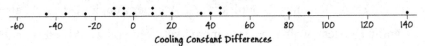
Cooling Constant Differences

Twelve of the 19 mice had cooling constants that were greater when they were freshly killed than when they were reheated. The dotplot of differences appears to have at least one extreme outlier at 130. The mean difference in cooling constants was 20.8, the standard deviation was 45.8.

c. Let μ_d represent the mean of the population of differences in cooling constants of the mice when they are freshly killed minus when they are reheated.

H_0: $\mu_d = 0$ vs. H_a: $\mu_d \neq 0$.

Technical conditions: You have no indication that these mice were randomly selected from the population of all mice. The sample size is small $(19 < 30)$, and the right skewness in the data and the extreme outliter indicate these data may not have come from a normal population of differences. You should probably stop the analysis at this point, or you could consider the analysis approximate as the skewness is not severe and 19 is a moderate sample size. But you should interpret the results below with caution.

The test statistic is $t = \dfrac{20.8}{45.8 / \sqrt{19}} = 1.98.$

Using Table III with 18 degrees of freedom, $1.734 < 1.98 < 2.1010$, so $2 \times .025 < $ p-value $ < 2 \times .05$. This means $.05 < $ p-value $ < .10$. Using technology, the p-value is .064.

Do not reject H_0 at the $\alpha = .05$ level but do reject H_a at the $\alpha = .10$ level. You do have some statistical evidence to conclude that the cooling constants of freshly killed mice and reheated mice differ, on average (but remember that you did not meet the technical conditions).

d. A 95% CI for μ_d is $20.8 \pm t^*(10.51)$.

Using Table III with 18 degrees of freedom, $t^* = 2.1010$, so the confidence interval is $(-1.28, 42.88)$. Using Minitab, the interval is $(-1.3, 42.9)$. Using the applet, the interval is $(-1.275, 42.875)$.

If the technical conditions had been met, you would be 95% confident the cooling constant of freshly killed mice is somewhere between 1.3 smaller and 4.29 larger than that of reheated mice, on average. *Note*: A 90% confidence interval would not contain zero.

Exercise 23-19: Exam Score Improvements

a. The following dotplot shows the improvement in scores from exam 1 to exam 2:

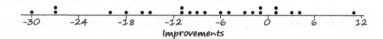

b. Student 19 had the greatest improvement in exam scores. He or she improved 11 points (from 84 to 95).

c. Student 8 had the greatest decline in exam scores. He or she declined 30 points (from 94 to 64).

d. Five of the 23 (or .2174) students scored higher on exam 1 than on exam 2.

e. The sample mean improvement is –8.7 points. The sample SD is 10.73 points.

f. Let μ_d represent the population mean improvement in the exam scores from exam 1 to exam 2.
The null hypothesis is that the population mean improvement from exam 1 to exam 2 is 0 (no improvement or decline in the mean scores). In symbols, $H_0: \mu_d = 0$.

The alternative hypothesis is that the population mean improvement from exam 1 to exam 2 differs from 0 (i.e., the mean score either improved or declined). In symbols, $H_a: \mu_d \neq 0$.

The test statistic is $t = \dfrac{-8.7}{10.73 / \sqrt{23}} = -3.89$

Using Table III with 22 degrees of freedom, $-3.89 < -3.792$, so the p-value is off the chart. This means p-value $< 2 \times .0005 = .001$. Using technology, the p-value is .001.

g. Yes, with this small p-value, reject the null hypothesis at the .10, .05 and .01 significance levels.

d. A 95% confidence interval for μ_d is (–13.34, –4.05).

You are 95% confident the score on exam 1 is between 4.05 and 13.34 points higher, on average, than the score on exam 2 for the population represented by these 23 students.

i. Technical conditions: The sample size is not large ($n = 23 < 30$), but a probability plot of the improvements indicates they could plausibly have come from a normally distributed population.

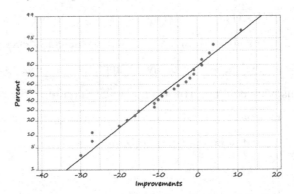

However, the scores were not gathered from a sample random sample, so you should be cautious in generalizing your results beyond this particular statistics class. For this class, you had very strong statistical evidence (p-value = .001) that the exam scores had declined, on average between 4 and 13.3 points from the first to the second exam (with 95% confidence). This is a decline of about half to more than a full letter grade, so it would be a *practically* significant decline in the exam scores.

Exercise 23-21: Exam Score Improvements

a. Let π represent the proportion of test takers who improved from exam 1 to exam 2.

Test: H_0: $\pi = .5$ vs. H_a: $\pi < .5$. Use a one-proportion z-test.

b. Technical conditions:

$n\pi_0 = n(1 - \pi_0) = 11.5 > 10$, so this technical condition is satisfied.

However, the data were not from a simple random sample of statistics students' exam scores, so you should be cautious about generalizing these results beyond this particular statistics class.

The test statistic is $z = \dfrac{.217 - .5}{\sqrt{\dfrac{(.5)(.5)}{23}}} = -2.71$.

p-value $= \Pr(Z < -2.71) = .0033$.

Because p-value $= .0033 < .01$, reject H_0 at the $\alpha = .01$ level. You have strong statistical evidence that the proportion of test takers in the population you are willing to generalize to from this statistics class who improved from exam 1 to exam 2 is less than one-half.

Exercise 23-23: Comparison Shopping

a. This would be a categorical variable; it would not have a numerical value.

b. Thirteen of the items cost more at Lucky's, 16 of the items cost more at Vons, and 5 of the items cost the same at both stores.

c. If you ignore the items that cost the same at both stores, 29 items remain: $13/29 \approx .448$ so about 45% of the items cost more at Lucky's.

d. Let π represent the proportion of all grocery items that cost more at Luckys than at Vons.

Test: H_0: $\pi = .5$ vs. H_a: $\pi \neq .5$. Use a one-proportion z-test.

d. Technical conditions:

$n\pi_0 = n(1 - \pi_0) = 14.5 > 10$, so this technical condition is satisfied.

You do not know, however, whether this is a simple random sample of items available at both stores, so you are unable to check this technical condition and should therefore be cautious about generalizing your conclusions to all items at all Lucky's and Vons stores in this town.

The test statistic is $z = \dfrac{.448 - .5}{\sqrt{\dfrac{(.5)(.5)}{29}}} = -0.56$.

p-value $= 2 \times \Pr(Z < -0.56) = 2 \times .2912 = .5824$.

Because the p-value is large, do not reject H_0 at any commonly used significance level. You have no real statistical evidence that the proportion of all grocery items that cost more at Lucky's than Vons is different from one-half.

e. For a 96% CI for π, you calculate $.448 \pm (2.05)(.093) = (.259, .638)$. You are 96% confident that between 25.9% and 63.8% of the grocery products cost more at Lucky's than at Vons (provided this is a random sample).

Exercise 23-25: Word Twist

a. The mean of the differences for your friend would be the opposite (negative) of the mean that you obtain. This makes sense because if you and your friend subtract in opposite orders, your lists of differences will be the same except for the signs of each value. One set of differences will be the other set multiplied by the value negative one.

b. The standard deviation of both sets of differences would be the same for you and your friend. The order in which you subtract does not affect how the differences are spread from the mean.

c. The test statistic you obtain will be the positive version of the test statistic that your friend obtains. All of the values used to compute the test statistic will be the same for both of you, except for the mean difference. This mean difference will have the opposite sign.

d. The p-value will be the same for both you and your friend because the t-distribution is perfectly symmetric, so $\Pr(T \geq t) = \Pr(T \leq -t)$.

e. If the endpoints of your confidence interval are (A, B), then the endpoints of your friend's interval will be $(-B, -A)$. Both intervals will have the same width, but your interval will be the "positive version" of your friend's interval.

Exercise 23-27: Muscle Fatigue

a. The null hypothesis is that the population mean female time until muscle fatigue is the same as the population mean male time until muscle fatigue. In symbols, $H_0: \mu_F = \mu_M$.

The alternative hypothesis is that the population mean female time until muscle fatigue is not the same as the population mean male time until muscle fatigue. In symbols, $H_a: \mu_F \neq \mu_M$.

Technical conditions: The sample sizes are small ($n_1 = n_2 = 10 < 30$). A probability plot indicates the male times might plausibly come from a normal distribution, but the population of female times is unlikely to be normal. You are assuming the sample data come from two independent samples, so this condition would be met.

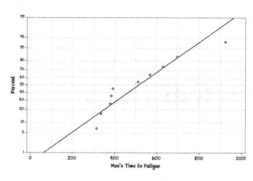 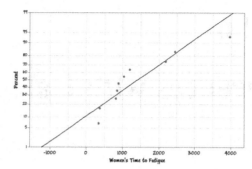

The test statistic is $t = \dfrac{1408 - 513}{\sqrt{\dfrac{(1133)^2}{10} + \dfrac{(194)^2}{10}}} = 2.46$.

Using Table III with 9 degrees of freedom, $2.262 < 2.46 < 2.281$, so $2 \times .01 < p\text{-value} < 2 \times .025$. This means $.02 < p\text{-value} < .05$. Using Minitab, the p-value is .036. Using the applet, the p-value is .0360.

Test decision: Because the p-value is less than .05, reject H_0 at the $\alpha = .05$ significance level.

Conclusion in context: You have moderate statistical evidence that the population mean female time until muscle fatigue is not the same as the population mean male time until muscle fatigue.

b. The p-value has not changed from the paired t-test.

c. Because the p-value is the same in both tests, the pairing was not particularly useful in this study.

Exercise 23-29: On Your Own

Answers will vary.

Unit 6

Inferences with Categorical Data

Topic 24

Goodness-of-Fit Tests

Odd- Numbered Exercise Solutions

Exercise 24-7: Mendel's Peas

a. The expected counts should not all be the same because the hypothesized proportions are not the same for each category.

b. When the *p*-value is large (greater than .2), you *do not* reject the null hypothesis.

c. The decision to reject or not reject the null hypothesis is based on the value of the *p*-value, not the test statistic.

d. The test statistic is calculated by dividing each (*observed – expected*) entry by the *expected count*, not by the observed count.

e. The expected counts here have all been rounded to the nearest integer, which can create a significant round-off error in the value of the test statistic.

Exercise 24-9: Leading Digits

a. The following bar graph displays the data:

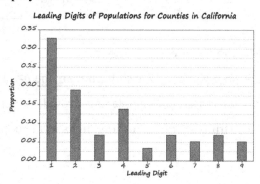

This distribution of leading digits is strongly skewed to the right with more than 30% of the leading digits being 1.

b. The expected counts are shown in the following table:

Leading Digit	1	2	3	4	5	6	7	8	9	Total
Expected Count (E)	17.458	10.208	7.25	5.626	4.582	3.886	3.364	2.958	2.668	58

c. No, all of the expected counts are not at least 5, so this technical condition is not met.

d. The new observed counts are shown in the table below:

Leading Digit	1	2	3	4	5 or 6	7, 8 or 9	Total
Observed Count (O)	19	11	4	8	6	10	58

e. Let π_i represent the probability of digit i in this process.

H_0: $\pi_1 = .301$, $\pi_2 = .176$, $\pi_3 = .125$, $\pi_4 = .097$, $\pi_{5,6} = .146$, $\pi_{7-9} = .155$

H_a: At least one of these probabilities is incorrect.

Technical conditions: The expected counts are all at least five (after collapsing the table), with the smallest equal to 5.626. These data are not a random sample of the populations in California counties because you have analyzed all California counties in 2006. However, you might consider the 2006 population counts as realizations of some overall process; you have shown that the distribution of leading digits is within random sampling variability of what you would expect based on Benford's model.

	1	2	3	4	5 or 6	7, 8 or 9	Total
Observed Count (O)	19	11	4	8	6	10	58
Expected Count (E)	17.458	10.208	7.25	5.626	8.468	8.99	58
$(O - E)^2/E$	.1362	.0615	1.4569	1.0018	.7193	.1135	3.489

The test statistic is $\chi^2 = 3.489$ with 5 degrees of freedom.

Using Table IV, $3.489 < 7.29$, so p-value $> .2$. Using Minitab, the p-value is 0.625.

With the large p-value, do not reject H_0. The sample data provide no reason to doubt that Benford's adjusted model describes the distribution of the leading digits of the populations of the 58 counties in California.

Exercise 24-11: Wayward Golf Balls

a. The following bar graph displays these results:

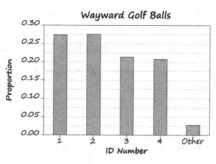

The ID numbers 1 and 2 seem somewhat more common than 3 and 4. The percentage of balls in the "other" category is very small (less than 4%).

b. Let π_i represent the probability of a ball having identification number i.

H_0: $\pi_1 = \pi_2 = \pi_3 = \pi_4 = \frac{1}{4}$.

H_a: At least one π_i is not $\frac{1}{4}$.

Each ID number would have an expected count of $.25 \times 486 = 121.5$.

c. The test statistics is $\chi^2 = 1.977 + 2.241 + 1.730 = 8.469$.

d. Using Table IV with 3 degrees of freedom, you find $7.81 < 8.469 < 9.35$, so $.025 < p\text{-value} < .05$. Using Minitab, the p-value is .0373.

e. You would reject the "equally likely" hypothesis at the $\alpha = .10$ level and at the $\alpha = .05$ level. You would *not* reject at the $\alpha = .01$ level.

f. If there were ten times as many balls in his yard with the same proportional breakdown of ID numbers, the test statistic should be ten times larger, and the p-value value should be smaller. You would expect to reject the null hypothesis of "equally likely" ID numbers at virtually any significance level. This makes sense because the larger the sample size, the more statistically significant the result, all other things being equal.

g. Now the expected counts for each ID number are 1215 (ten times the previous expected counts). The test statistic is $\chi^2 = 19.774 + 22.407 + 17.305 + 25.206 = 84.691$.

Using Table IV with 3 degrees of freedom, $84.691 > 17.73$, so the p-value $< .0005$.

h. Because these golf balls were used on the same golf course, and were found in the back yard of one statistics teacher, you should probably only generalize these results to the balls used by golfers at this course. You could probably consider these balls representative of the balls used on this course, and perhaps on other courses in this city.

Exercise 24-13: Halloween Treats

a. Let π_{candy} represent the proportion of all trick-or-treaters in this population who would choose the candy, and π_{toy} the population proportion who would choose the toy.

H_0: $\pi_{candy} = .5$, $\pi_{toy} = .5$.

H_a: $\pi_{candy} \neq .5$, $\pi_{toy} \neq .5$.

The expected count for both groups is $.5 \times 283 = 141.5$, so this technical condition is satisfied.

The test statistic is $\chi^2 = (148 - 141.5)^2/141.5 + (131 - 141.5)^2/141.5 = .2986 + .2986 = .5972$.

Using Table IV with 1 degree of freedom, $.5972 < 1.62$, so p-value $> .2$.

Using Minitab, the p-value is .440.

Because the p-value is not small, do not reject H_0 at the $\alpha = .2$ significance level.

You have no real statistical evidence that H_0 is false. That is, you have no reason to doubt that the population of trick-or-treaters is equally likely to choose the toy and the candy.

Technical conditions: The expected counts are at least five in both categories. However, this is not a random sample of trick-or-treaters. All the households were in five suburban Connecticut neighborhoods, so you should be very cautious in generalizing your results beyond this population.

b. You calculate $\dfrac{135}{283} \pm (1.28)\sqrt{\dfrac{(.477)(.523)}{283}} = .477 \pm .038 = (.439, .515)$.

You are 80% confident the population percentage of trick-or-treaters who would choose the toy over the candy is between 43.9% and 51.5%.

c. This interval does include the value .5. This is consistent with your test decision in part a because the test decided that .5 was a plausible value for the proportion of trick-or-treaters who would choose the toy at the .2 significance level.

d. Because this was not a random sample of trick-or-treaters and all the households were in one of five suburban Connecticut neighborhoods, you probably should not generalize your results beyond suburban Connecticut neighborhoods.

Exercise 24-15: Candy Colors

a. The following bar graph displays the data:

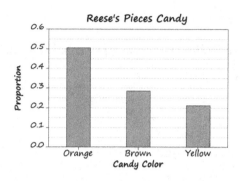

It appears that brown and yellow are equally likely colors, but orange is much more likely to occur than either of them. Almost half of the sample was orange, whereas between 21%–28% was yellow or brown.

c. Let π_{orange} represent the proportion of all Reese's Pieces that are orange.

H_0: $\pi_{orange} = .5$, $\pi_{brown} = \pi_{yellow} = .25$.

H_a: At least one of these proportions is incorrect.

Technical conditions: The expected counts in each category are at least five. You were not told whether this was a random sample of Reese's pieces, but it was very likely to be a representative sample even if it were not randomly selected.

	Orange	Brown	Yellow	Total
Observed Count	273	154	114	541
Expected Count (E)	270.5	132.25	135.25	541
$(O - E)^2/E$	.0231	2.599	3.387	5.961

The test statistic is $\chi^2 = 5.961$.

Using Table IV with 2 degrees of freedom, $4.61 < 5.961 < 5.99$, so $.05 < p\text{-value} < .10$. Using Minitab, the p-value is .0508.

Because the p-value is greater than .05, do not reject H_0 at the .05 significance level.

You do not have enough statistical evidence to conclude that the population percentages of colors differ from 50% orange, 25% brown, and 25% yellow at the 5% level of significance (although you would at the 10% level of significance).

d. Yes, there are probably many null hypotheses about the color distribution that would not be rejected at the .05 significance level. For example, $\pi_{orange} = .5$, $\pi_{brown} = .26$, $\pi_{yellow} = .24$, or $\pi_{orange} = .51$, $\pi_{brown} = .25$, $\pi_{yellow} = .24$ would probably not be rejected.

Exercise 24-17: Flat Tires

a. Answers will vary by class. The following is one representative set of answers.

Here is a bar graph of the class results:

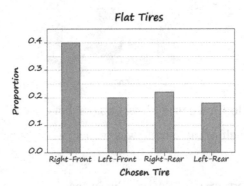

The right-front tire is a much more frequent choice than the others in this class. It was chosen about 40% of the time, whereas the other tires were each chosen roughly 20% of the time.

b. Let $\pi_{\text{right-front}}$ represent the probability of a person in this population choosing the right-front tire.

H_0: $\pi_{\text{right-front}} = \pi_{\text{left-front}} = \pi_{\text{right-rear}} = \pi_{\text{left-rear}} = ¼ = .25$.

H_a: At least one of these proportions is not .25.

Technical conditions: The expected counts in each category are at least five (see table below). This is not a random sample of students from this school as the sample consists of the students in one class, but it is likely to be a representative sample for this school (there is no reason to suspect students in this class would answer this question differently from students in other classes).

	Right-Front	Left-Front	Right-Rear	Left-Rear	Total
Observed Count	20	10	11	9	50
Expected Count (E)	12.5	12.5	12.5	12.5	50
$(O - E)^2/E$	4.5	.5	.18	.98	6.16

The test statistic is $\chi^2 = 6.16$.

Using Table IV with 3 degrees of freedom, $4.64 < 6.16 < 6.25$, so $.1 < p$-value $< .2$. Using Minitab, the p-value is .104.

c. H_0: $\pi_{\text{right-front}} = .40$, $\pi_{\text{left-front}} = \pi_{\text{right-rear}} = \pi_{\text{left-rear}} = .20$.

H_a: At least one of these proportions is incorrect.

	Right-Front	Left-Front	Right-Rear	Left-Rear	Total
Observed Count	20	10	11	9	50
Expected Count (E)	20	10	10	10	50
$(O - E)^2/E$	0	0	.1	.1	.2

The test statistic is $\chi^2 = 0.2$.

Using Table IV with 3 degrees of freedom, $.2 < 4.64$, so p-value $> .2$. Using Minitab, the p-value is .978.

With such a large p-value, do not reject H_0 at the .05 significance level. You do not have sufficient statistical evidence to suggest that, at this school, the population proportions of tire choices differ from $\pi_{\text{right-front}} = .40$; $\pi_{\text{left-front}} = \pi_{\text{right-rear}} = \pi_{\text{left-rear}} = .20$ in this population (assuming a representative sample).

Exercise 24-19: Calling Heads or Tails

a. Answers will vary. Here is one representative set.

Let π_{heads} represent the probability of a student at this school calling "heads".

H_0: $\pi_{\text{heads}} = .7$, $\pi_{\text{tails}} = .3$. H_a: $\pi_{\text{heads}} \neq .7$, $\pi_{\text{tails}} \neq .3$.

Technical conditions: The expected counts in each category are at least five (see table below). This is not a random sample of students at this school, but it is probably a representative sample as there may not be any reason to suspect students in this class will call "heads" at a different rate than students at this school in general.

	Heads	Tails	Total
Observed Count	16	4	20
Expected Count (E)	14	6	20
$(O - E)^2/E$	.2857	.667	.9254

The test statistic is $\chi^2 = .9524$.

Using Table IV with 1 degree of freedom, $.9524 < 1.64$, so p-value $> .2$. Using Minitab, the p-value is .329.

b. With a *p*-value of .329 > .10, you would not reject the null hypothesis at the α = .10 significance level. There is no reason to doubt that 70% of the students at this school would call "heads" in this situation (assuming a representative sample).

c. Here are the expected counts:

	Heads	Tails	Total
Observed Count	32	8	40
Expected Count (E)	28	12	40
$(O - E)^2/E$	.5714	1.333	1.905

The test statistic is $\chi^2 = 1.905$.

Using Table IV with 1 degree of freedom, $.1 < p\text{-value} < .2$. Using Minitab, the *p*-value is .168. You would still not reject the null hypothesis at the α = .10 significance level. There is no reason to doubt that 70% of the students at this school would call "heads" in this situation.

d. The expected counts and test statistic doubled, and the *p*-value was cut in half when the sample size was doubled.

Exercise 24-21: Water Taste Test

a. Here is the bar graph:

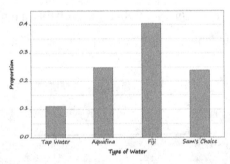

This bar graph indicates that students who like each type of water is not the same. The Fiji water was a clear favorite with 40% of the students perferring this type. Only about 11% of the students indicated that they preferred the tap water, and roughly 25% of the students preferred either Aquafina or Sam's Choice.

b. Let $\pi_{\text{tap water}}$ represent the population proportion of students who prefer the tap water, etc.

H_0: $\pi_{\text{tap water}} = \pi_{\text{Aquafina}} = \pi_{\text{Fiji}} = \pi_{\text{Sam's Choice}} = \frac{1}{4} = .25$.

H_a: At least one of these proportions is not .25.

Technical conditions: The expected counts in each category are at least five (see table below). This is not a random sample of students at this school, and it may not be representative because all the students in this study were attending Oktoberfest. So you should be cautious about extending the results to a larger population.

	Tap Water	Aquafina	Fiji	Sam's Choice	Total
Observed Count	12	27	44	26	109
Expected Count (E)	27.25	27.25	27.25	27.25	109
$(O-E)^2/E$	8.534	0.0023	10.296	0.057	18.890

The test statistic is $\chi^2 = 18.890$.

Using Table IV with 3 degrees of freedom, $18.89 > 17.73$, so p-value $< .0005$. Using Minitab, the p-value $\approx .0000$.

With such a small p-value, you reject the null hypothesis at any commonly used significance level. You have found very strong statistical evidence that all four types of water are not equally preferred.

c. The Fiji water makes the largest contribution to the chi-square test statistic. Its observed count (44) is much greater than its expected count (27.25) which indicates that the students preferred the Fiji water more than you would expect if all four types of water were equally preferred. The tap water has the next largest contribution to the test statistic, and its observed count (12) is less than half of its expected count (27.25). This indicates that less than a quarter of the population prefers tap water.

d. H_0: $\pi_{tap\ water} = .1429$, $\pi_{Aquafina} = \pi_{Fiji} = \pi_{Sam's\ Choice} = .2857$.

H_a: At least one of these proportions is as given in the null hypothesis.

	Tap Water	Aquafina	Fiji	Sam's Choice	Total
Observed Count	12	27	44	26	109
Expected Count (E)	14.29	28.57	28.57	28.57	109
$(O-E)^2/E$	.00753	.00505	.0487	.00789	.06908

The test statistic is $\chi^2 = .06908$.

Using Table IV with 3 degrees of freedom, $.06908 < 4.64$, so the p-value $> .2$. Using Minitab, the p-value is .995.

With such a large p-value, you do not reject the null hypothesis at any commonly used significance level.

You have found no statistical evidence to doubt that the population proportions who prefer the tap water is half as large as the population proportions who prefer each of the three brands of bottled water, and that those brands of bottled water are equally preferred.

Exercise 24-23: Water Taste Test

Let $\pi_{\text{tap water}}$ represent the population proportion of students who least prefer the tap water, etc.

H_0: $\pi_{\text{tap water}} = .5$, $\pi_{\text{Aquafina}} = \pi_{\text{Fiji}} = \pi_{\text{Sam's Choice}} = 1/6$.

H_a: At least one of these proportions is not correct.

Technical conditions: The expected counts in each category are at least five (see table below). This is not a random sample of students at this school, and it may not be representative because all the students in this study were attending Oktoberfest. So you should be cautious about extending the results to a larger population.

	Tap Water	Aquafina	Fiji	Sam's Choice	Total
Observed Count	51	18	17	21	107
Expected Count (E)	53.5	17.833	17.833	17.833	107
$(O - E)^2/E$	.00157	0.0390	0.562	0.117	0.719

The test statistic is $\chi^2 = 0.719$.

Using Table IV with 3 degrees of freedom, $.719 < 4.64$, so p-value $> .2$. Using Minitab, p-value $= .869$.

With the large p-value, do not reject H_0 at any commonly used significance level.

You have no statistical evidence that would cause you to doubt that the population proportion who least prefer tap water is .5, and that population proportions who least prefer each of the bottled waters is one-sixth.

Exercise 24-25: Suicides

a. Note that the expected counts are all the same (36). Each of these values was obtained by multiplying the total number of observations (252) by 1/(number of categories) = 1/7. So you calculate $252 \times 1/7 = 36$.

b. The chi-square contribution for Monday was calculated as $(30 - 36)^2/36 = 1.00$.

c. The p-value (.001) is the probability of obtaining sample counts that result in a chi-square test statistic as large as 23.7778 or larger, assuming that the probability of a suicide on each of the seven week days is the same.

d. Because the *p*-value is so small (.001), you would reject the null hypothesis at any commonly used significance level. You have very strong statistical evidence that the probability of a suicide occurring on any of the seven weeks days is not the same. At least one of the seven days is more or less likely to have a suicide than the others.

e. The data *do not* appear to support the idea that Mondays are likelier days for suicides. The day that made the largest contribution to the test statistic is Thursday (18.7778), and its expected count (36) was much smaller than its observed count (62). This indicates that a suicide is more likely to occur on a Thursday than on any other day.

f. If the sample sizes were larger, but the sample percentages stayed the same, the *p*-value would decrease (the results would become more statistically significant).

Exercise 24-27: Rookie Quarterbacks

a. The observational units are the readers of *Sports Illustrated* who voted on their website.

The variable is *which of the four quarterbacks does the reader think would have the best career*.

b. Here is a bar graph of the data:

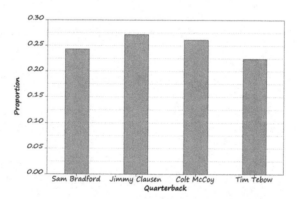

These data indicate that the differences in sample proportions among these four quarterback choices were not very large. Although Jimmy Clausen was the most popular choice (27.4%), the least popular choice was still selected more than 22% of the times.

c. To compute the *p*-value, you would use 3 degrees of freedom.

Using Table IV, 17.73 < 246.8 (test statistic), so *p*-value < .0005.

Using Minitab, *p*-value ≈ .0000.

d. The *p*-value is so small (even though the sample proportions are fairly similar) because the sample size is extremely large (46,377). With such a large sample size, even a very small difference in proportions can be statistically significant.

e. The sample data do provide strong statistical evidence that the four quarterbacks are not equally likely to be selected by this population. The *p*-value is very small, and so the null hypothesis would be rejected at any commonly used significance level. If the all four quarterbacks were equally likely to be selected by this population, you would virtually never find sample proportions that result in a chi-square test statistic as large as 246.8.

Exercise 24-29: On Your Own

Answers will vary.

Topic 25

Inference for Two-Way Tables

Odd- Numbered Exercise Solutions

Exercise 25-7: Pursuit of Happiness

a. The data were collected via random sampling.

b. Here is the segmented bar graph:

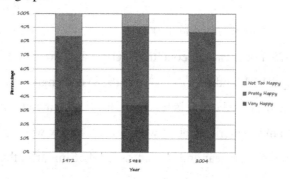

This graph reveals does not reveal a very strong association between year and happiness level. In other words, the distributions of happiness level appear fairly similar among the three years. The highest percentage of "very happy" subjects (34%) did occur in 1988, and this year also had the smallest percentage of "not too happy" subjects (9.3%).

c. The null hypothesis is H_0: *Year* and *happiness level* are independent variables.

The alternative hypothesis is H_a: There is an association between *year* and *happiness level*.

The expected counts (E) and observed counts are shown in the table below:

	1972	1988	2004	Total
Very Happy (O)	486	498	419	
Expected (E)	511.05	466.50	425.45	1403
$(O - E)^2/E$	1.228	2.127	0.098	
Pretty Happy (O)	855	832	738	
Expected (E)	883.32	806.32	735.37	2425
	0.908	0.818	0.009	

$(O - E)^2/E$				
Not Too Happy (O) **Expected (E)** $(O - E)^2/E$	265 211.63 13.458	136 193.18 16.927	180 176.18 0.083	581
Total	1606	1466	1337	4409

All of the expected counts are larger than five. The subjects were randomly selected from the population of adult Americans, so the technical conditions are satisfied.

Using Minitab with 4 degrees of freedom, the chi-square test statistic is $\chi^2 = 35.655$ and the p-value $\approx$ .000.

Using Table IV, $35.655 > 20.00$, so p-value $< .0005$.

Because the p-value $< .075$, reject H_0 and conclude there is an association between year and happiness level. The (not too happy, 1972) and (not too happy, 1988) cells make the largest contributions to the test statistic. Because the observed count is less than the expected count in 1988, you know the population proportion of adult Americans who were not too happy was less than you would expect if there were no association between year and happiness level. Similarly, the population proportion of adult Americans who were not too happy in 1972 was greater than you would expect if there were no association between the variables.

Exercise 25-9: Baldness and Heart Disease

a. This is an observational study because the researchers did not randomly assign men to the baldness groups.

b. The expected counts in each cell would not all be 5. In particular, in the extreme baldness column, the expected counts would be 1.39 (heart disease) and 1.61 (control).

c. Here is the 2 × 4 table:

	None	Little	Some	Much or More
Heart Disease	251	165	195	52
Control	331	221	185	35

d. H_0: *Degree of baldness* and *heart disease* are independent variables for males in this population.

   H_a: *Degree of baldness* and *heart disease* are related.

Using Minitab, the test statistic is $\chi^2 = 14.510$ with 3 degrees of freedom, and the *p*-value is .002.

Using Table IV, $12.84 < 14.51 < 16.27$, so $.001 < p\text{-value} < .005$.

With such a small *p*-value, reject H_0 at any commonly used significance level.

e. You have strong statistical evidence that degree of baldness and heart disease are related for males in this population. However, because this is an observational study, you cannot conclude that baldness *causes* heart disease, only that there is an association between these variables in this population.

Exercise 25-11: Asleep at the Wheel

a. This is an observational study because the researchers did not randomly assign the subjects to the crash or control groups.

b. Here is the segmented bar graph:

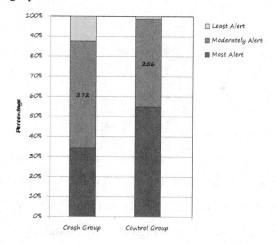

This graph reveals that group (crash or control) and sleepiness appear to be related. More than 54% of the control group was classified as "most alert," whereas only 34% of the crash group was classified this way. More than 10% of the crash group was classified as "least alert," but less than 2% of the control group was.

c. H_0: *Crash group* and *sleepiness classification* are independent variables among New Zealand drivers.

H_a: There is an association between *crash group* and *sleepiness classification* in this population.

Technical conditions: The expected counts are at least five in each cell of the table. You do not know whether this was a random sample of all New Zealand drivers.

Using Minitab, the test statistic is $\chi^2 = 81.692$ with 2 degrees of freedom and the *p*-value $\approx .000$.

Using Table IV, $81.692 > 15.20$, so *p*-value $< .0005$.

With such a small *p*-value, reject H_0 and conclude that whether a driver crashes is related to his/her level of sleepiness.

d. You have very strong statistical evidence that there is an association between sleepiness and involvement in car crashes among New Zealand drivers, assuming these samples are representative. However, because this is an observational study, you cannot conclude from this study that sleepiness *causes* car crashes.

Exercise 25-13: Asleep at the Wheel

Here is a segmented bar graph of these data:

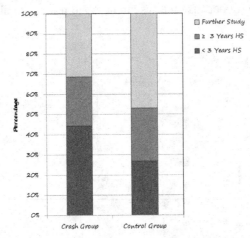

The graph indicates an association between *education level* and *car crashes* for subjects in the sample. Forty-four percent of those involved in crashes had less than 3 years of high school whereas 47% of those in the control group had studied beyond high school. About 25% of both groups had at least 3 years of high school but no further study.

H_0: *Car crashes* and *education level* are independent variables in the population of New Zealand drivers.

H_a: There is an association between *car crashes* and *education level* in this population,

Technical conditions: The expected counts are at least five for each cell of the table. You do not know whether this was a random sample of New Zealand drivers.

Using Minitab, the test statistic is $\chi^2 = 43.436$ with 2 degrees of freedom and *p*-value $\approx .000$. Using Table IV, $43.436 > 15.20$, so *p*-value $< .0005$.

Test decision: With such a small *p*-value, reject H_0 at any commonly used significance level.

Conclusion in context: You have very strong statistical evidence that there is an association between education level and involvement in car crashes among New Zealand drivers, assuming this sample is representative. However, because this was an observational study, you cannot conclude from this study that using alcohol *causes* car crashes.

Exercise 25-15: Weighty Feelings

a. The data collection scenario used in this study was random sampling from one population. The researchers classified the respondents both by gender and their feelings about their weight after they were randomly selected.

b. The following segmented bar graph compares the distribution of weight self-images between men and women:

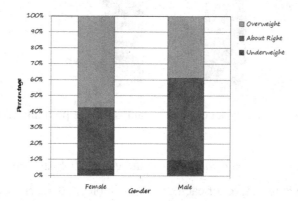

Males in this study were almost 3 times as likely as females to feel underweight (.096 vs. .038). More than half of the females (.573) felt overweight, whereas more than half of the males (.515) felt about right about their weight.

c. H_0: The distribution of weight self-images is the same between males and females (*gender* and *weight self-image* are independent variables in the population of adult Americans).

   H_a: The distribution of weight self-images is not the same between males and females.

   Technical conditions: The expected counts are at least five for each cell of the table, and this was a randomly selected sample.

   Using Minitab, the test statistic is $\chi^2 = 226.579$ with 2 degrees of freedom and *p*-value $\approx .000$. Using Table IV, $226.579 > 15.20$, so *p*-value $< .0005$.

   With such a small *p*-value, reject H_0 at any commonly used significance level. You have very strong statistical evidence that the distribution of weight self-images is not the same between males and females in the population.

Exercise 25-17: College Students' Drinking

a. This is an observational study because the researchers did not assign students to their genders; these were predetermined.

b. Here is the segmented bar graph:

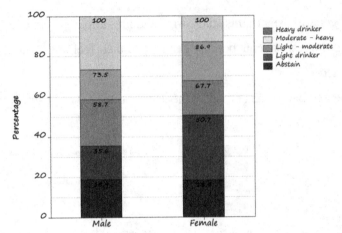

c. H_0: *Drinking level* and *gender* are independent variables for the population of students at Rutgers University

H_a: There is an association between *drinking level* and *gender* for this population.

d. The expected counts are at least five for each cell of the table, and observations arise from a random sample from the population, so the technical conditions are met.

e. Because the *p*-value is small ($< .001$), reject H_0 at any commonly used significance level. You have very strong statistical evidence that there is association between drinking level and gender for the population of Rutgers University students.

f. The (male, light drinker) cell contributes the most to the test statistic. Its observed count (44) is much less than its expected count (67.21). The (male, heavy drinker) makes the next largest contribution to the test statistic, and its observed count (70) is greater than its expected count (50.08). For each of these drinking categories, the female cells also make large contributions to the test statistic, but the relationships between observed and expected counts in these categories are reversed from the male cells. This indicates that there are more male heavy drinkers, and fewer male light drinkers than you would expect if there were no association between gender and drinking level. Similarly, there are fewer female heavy drinkers and more female light drinkers that you would expect if there were no association between the variables. (In other words, males are more likely to be heavy drinkers than are the females, and the males are less likely to be light drinkers than are the females.)

Exercise 25-19: Suitability for Politics

a. The appropriate chi-square test for these data is independence because the researchers took one random sample and then classified the subjects by *both* explanatory and response variables.

b. H_0: *Political inclination* and *reaction to the statement* are independent variables in the population of adult Americans.

H_a: *Political inclination* and *reaction to the statement* are associated variables in this population.

Here is a segmented bar graph displaying these data:

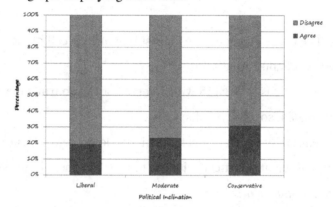

Technical conditions: The expected counts are at least five in each cell of the table, and this was a randomly selected sample of adult Americans.

Using Minitab, the test statistic is $\chi^2 = 9.867$ with 2 degrees of freedom and the *p*-value is .007. Using Table IV, $9.21 < 9.867 < 10.60$, so $.005 < p\text{-value} < .010$.

With the small *p*-value ($.007 < .05$), reject H_0 at any commonly used significance level.

You have strong statistical evidence that political inclination and reaction to the statement are associated in the population of adult Americans. The largest contribution to the test statistic is made by the "conservative/agree" cell, and in this cell the observed count (96) is greater than the expected count (78.13). The next largest contribution is made by the "liberal/agree" cell, in which the observed count (40) is less than the expected count (52.26).

c. H_0: *Gender and reaction to the statement* are independent variables in the population of adult Americans.

H_a: *Gender and reaction to the statement* are associated variables in this population.

Here is a segmented bar graph displaying these data:

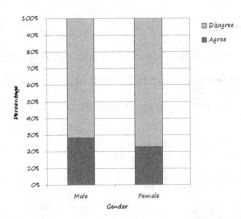

Technical conditions: The expected counts are at least five in each cell of the table, and this was a randomly selected sample of adult Americans.

Using Minitab, the test statistic is $\chi^2 = 3.155$ with 1 degree of freedom and the p-value is .076. Using Table IV, $2.71 < 3.155 < 3/84$, so $.05 < p$-value $< .10$.

Because the p-value equals .076, which is greater than .05, do not reject H_0 at the .05 significance level. You do not have sufficient statistical evidence to conclude there is an association between *gender* and *reaction to the statement*.

d. The p-value is smaller for the political inclination test. In this case, this means there is strong evidence of an association between *political inclination* and *reaction to the statement* in the population, but not enough evidence to conclude there is an association between *gender* and *reaction to the statement*.

Exercise 25-21: Cold Attitudes

a. This was an observational study because the researchers observed and categorized the subjects' emotional states.

b. Here is the two-way table for positive emotions:

	High	Medium	Low	Total
Got Cold	21	29	37	87
Didn't Get Cold	90	82	75	247
Total	111	111	112	334

c. Let π_{low} represent the probability that a subject with low positive emotions develops a cold, and so on.

H_0: $\pi_{low} = \pi_{medium} = \pi_{high}$.

H_a: At least one of these population probabilities is not equal to the others.

Technical conditions: The expected counts are all at least five. You have no indication that these subjects were randomly selected. In fact, these were all "healthy volunteers," so they may not be representative of the population of adults. You should be cautious about generalizing your results beyond the population from which these volunteers were drawn.

Using Minitab, the test statistic is $\chi^2 = 5.76$ with 2 degrees of freedom and the p-value is .056. Using Table IV, $4.61 < 5.76 < 5.99$, so $.05 < p$-value $< .10$.

Because the p-value is $.056 < .075$, reject H_0 at the .075 significance level. You have statistical evidence that the population proportions are not equal. However, you cannot conclude the development of a cold is caused by one's positive emotional state because this was an observational study, not an experiment, and there may be many confounding variables for which you could not control.

d. Here is the two-way table for negative emotions:

	High	Medium	Low	Total
Got Cold	29	28	31	88
Didn't Get Cold	82	83	81	246
Total	111	111	112	334

H_0: $\pi_{low} = \pi_{medium} = \pi_{high}$.

H_a: At least one of these population probabilities is not equal to the others.

Technical conditions: The expected counts are all at least five. You have no indication that these subjects were randomly selected. In fact, these were all "healthy volunteers," so they may not be representative of the population of adults. You should be cautious about generalizing your results beyond the population from which these volunteers were drawn.

Using Minitab, the test statistic is $\chi^2 = 0.177$ with 2 degrees of freedom and the p-value is .915. Using Table IV, $.177 < 3.22$, so the p-value $> .2$.

Because the p-value is $.915 > .075$, do not reject H_0 at the .075 significance level. You have no statistical evidence that the population proportions are not equal. Even if you had concluded the population proportions differed, you could not conclude the development of a cold is caused by one's negative emotional state because this was an observational study, not an experiment. There are many confounding variables for which you could not control.

Exercise 25-23: A Nurse Accused

a. H_0: The population death rates on the Gilbert shifts and non-Gilbert shifts are the same (or, the probability of a patient dying on Gilbert's shift is the same as the probability of a patient dying on a shift that Gilbert is not working).

H_a: The population death rates on the Gilbert and non-Gilbert shifts are not the same.

	Gilbert Shifts	Other Shifts	Total
Patient Died	40	34	74
(Expected Count)	(11.59)	(62.41)	
No Patient Died	217	1350	1567
(Expected Count)	(245.41)	(1321.59)	
Total	257	1384	1641

Using Minitab, the test statistic is $\chi^2 = 86.48$ with 1 degree of freedom and p-value $\approx .000$. Using Table IV, $85.48 > 12.12$, so p-value $< .0005$.

With such a small p-value, you can reject H_0 at the .001 significance level. The difference in the death rates on the two types of shifts is statistically significant at the $\alpha = .001$ significance level.

b. If there were no difference in the death rates on the two types of shifts (Gilbert and non-Gilbert), you would virtually never see a result this extreme or more extreme, by random chance alone. Because you did see such an extreme result (which is virtually impossible to have occurred by random chance), you conclude there is a statistically significant difference in the death rates on the two types of shifts.

c. A defense attorney *could* reasonably argue that the higher death rate on Gilbert's shifts is due to a confounding variable because this is an observational study, not a randomized experiment.

d. A defense attorney *could not* reasonably argue that the higher death rate on Gilbert's shifts is due to random chance. The probability of the higher death rates being due to random chance is zero (see part b).

Exercise 25-25: Exciting Life

a. Because there are so few observations in the "other" category, and it does not make sense to combine it with any of the three other categories, you could just delete it from the table. (However, the expected counts in the "other" cells are both at least five, so you do not need to delete it for this reason.)

b. H_0: There is no association between *gender* and *feelings about life* for the population of all American adults in 2008.

H_a: There is an association between *gender* and *feelings about life* for this population.

Technical conditions: All of the expected counts are at least five, and the sample was randomly selected from the population of all American adults.

Using Minitab, the test statistic is $\chi^2 = 4.326$ with 2 degrees of freedom and the *p*-value is .115 (after deleting the "other" category).

Using Table IV, $3.22 < 4.326 < 4.61$, so $.1 < p\text{-value} < .2$.

You do not have sufficient statistical evidence to conclude there is any association between *gender* and *feelings about one's life* (exciting, routine, or dull) in the population of American adults in 2008.

c. Here are the observed counts for the collapsed table:

	Male	Female	Total
Exciting	300	347	647
Not Exciting	310	392	792
Total	610	739	1349

H_0: There is an association between *gender* and *feeling that one's life is exciting* for the population of American adults in 2008.

H_a: There is no association between *gender* and *feeling that one's life is exciting* for this population.

Using Minitab with 1 degree of freedom, the test statistic is $\chi^2 = .663$ and the *p*-value is .416.

Using Table IV, $.416 < 1.64$, so *p*-value $> .2$.

Because the *p*-value is large, do not reject the null hypothesis at any commonly used significance level. You have no evidence of an association between gender and feeling that one's life is exciting for the population of adult Americans in 2008.

In Exercise 21-34, the test statistic was $z = 0.81$, and this is roughly the square root of the χ^2 test statistic ($.81^2 = .656 \approx .66$). The *p*-values and conclusions are the same in both tests.

Exercise 25-27: Acupuncture Effectiveness

a. Here is the 2 × 3 table displaying the observed counts in this study:

	Real Acupuncture	Sham Acupuncture	Conventional Treatment
Improved	184	171	106
Did not improve	203	216	281

Here is a segmented bar graph displaying the data:

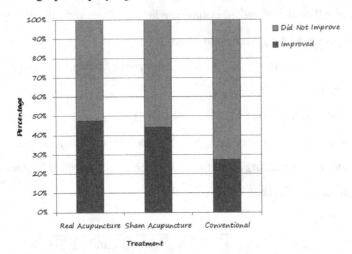

b. H$_0$: The population proportions of patients suffering from chronic low back pain who would improve were the same for all three treatment groups.

H$_a$: At least one of these population proportions differs from the rest.

Technical conditions: All of the expected counts for each cell in the table are at least five. The patients were not randomly selected from all people who suffer from chronic low back pain, or even from all Germans who suffer from chronic low back pain, so you should be cautious about extending these results to a larger population.

Using Minitab, the test statistic is $\chi^2 = 37.697$ with 2 degrees of freedom and the p-value $\approx .000$. Using Table IV, $37.697 > 15.20$, so p-value $< .0005$.

With this small p-value, reject H$_0$ at any commonly used significance level. You have found very strong statistical evidence that that proportion of German patients suffering from chronic low back pain who would improve is not the same for all three treatment groups. The conventional treatment group cells make the largest contributions to the test statistic, indicating that there were fewer improvements among the conventional treatment subjects than you would expect if the population proportions among all treatments were identical.

This was a well-designed experiment, so you can conclude that the treatment caused the differences in the proportions. However, because the sample was not randomly selected, you should not extend these results beyond patients in this study.

c. H_0: The population proportions of of patients suffering from chronic low back pain who would improve are the same for both the real and sham acupunture treatments.

H_a: The population proportions of improvements are the same for both treatment groups.

Here is a table of observed counts:

	Real Acupuncture	Sham Acupuncture	Total
Improved	184	171	355
Did not improve	203	216	419
Total	387	387	774

Using Minitab, the test statistic is $\chi^2 = .879$ with 1 degree of freedom and the *p*-value is .348. Using Table IV, .879 < 1.64, so *p*-value > .2.

Because the *p*-value > .2, you have no statististical evidence that the population success proportions for the real and sham acupuncure differ significantly at the .05 significance level.

Exercise 25-29: Coffee Consumption

Here is a segmented bar graph of these data:

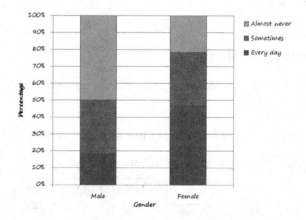

This graph indicates *gender* and *coffee consumption frequency* are not independent in this sample. Roughly 47% of the females in this sample reported drinking coffee every day, but only about 20% reported drinking coffee almost never. About half of the males reported drinking coffee almost never, and less than 20% of the males reported drinking coffee every day.

H_0: *Gender* and *coffee consumption frequency* are independent variables for this population.

H_a: There is an association between *gender* and *coffee consumption frequency* for the population of statistics students at this university.

Technical conditions: The expected counts for each cell in the table are at least five. However, these subjects were not randomly selected from all college students or even from all students at this university, so you should be cautious in generalizing these results beyond statistics students at this particular university.

Using Minitab with 2 degrees of freedom, the test statistic is $\chi^2 = 8.175$ and the p-value is .017. Using Table IV, $7.82 < 8.715 < 9.21$, so $.01 < p$-value $< .02$.

Because the p-value $= .017 < .05$, reject H_0 at the $\alpha = .05$ significance level. You have moderately strong statistical evidence that there is an association between *gender* and *coffee consumption frequency* for this population. The (female, every day) cell makes the largest contribution to the test statistic, and its observed value is greater than its expected value. This indicates that females are more likely to drink coffee every day than males. Similarly, an examination of the remaining contributions to the test statistic indicates that males are more likely to report drinking coffee almost never than females.

Because this was an observational study, you cannot conclude that gender is the cause of the difference in coffee consumption frequency between the males and females. Because the sample was not randomly selected from all students at this university, you should not extend these results beyond statistic students at this school.

Exercise 25-31: Feeling Rushed?

a. You calculate $1180 \times 977 / 4281 = 269.30$. This expected count is smaller than the observed count of 304.

b. For the upper right cell, the value $(O - E)^2/E$ is $(304 - 269.3)^2/269.3 = 4.472$.

c. The contributions to the test statistic are shown in the table below:

	1982	1996	2004
Always	9.186	2.850	4.472
Sometimes	0.001	0.004	0.012
Almost Never	12.821	3.504	7.011

d. Using Minitab, the test statistic is $\chi^2 = 39.861$ with 4 degrees of freedom and the p-value $\approx .000$.

e. Because the p-value $< .05$, reject the null hypothesis at the $\alpha = .05$ significance level. You have very strong statistical evidence that the population distributions of feelings about being rushed are not the same for all three years.

f. The ("almost never", 1982) cell contributes the most to the calculation of the test statistic. For this cell, the expected count (373.77) is smaller than the observed count (443), indicating more of the population almost never felt rushed in 1982 than you would expect if the population proportions were the same in all three years.

Exercise 25-33: On Your Own

Answers will vary.

Unit 7
Relationships in Data

Topic 26

Graphical Displays of Data

Odd- Numbered Exercise Solutions

Exercise 26-9: Challenger Disaster

a. A scatterplot of *O-ring failures* vs. *outside temperature* follows:

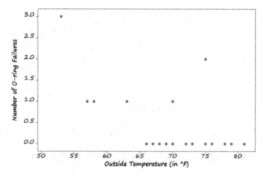

b. This scatterplot reveals a negative association between the *number of O-ring failures* and *outside temperature*. As the outside temperature rises, the number of O-ring failures tends to drop. All of the flights with no O-ring failure occurred when the outside temperature was at more than 65°F.

c. With a forecasted low temperature of 31°F, it seems very likely that there would be at least one O-ring failure.

d. The scatterplot of *O-ring failures* vs. *outside temperature* for the remaining seven flights follows:

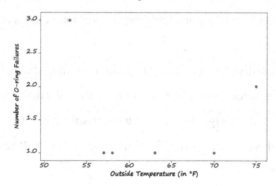

This scatterplot does not seem to reveal either a strictly positive or a strictly negative association between the variables. The relationship appears more parabolic. This scatterplot definitely does not make the case for a negative association as strongly as the previous scatterplot did.

e. The flights with no O-ring failures gave a great deal of information about how the number of failures was related to outside temperature. They should not have been excluded from the analysis.

Exercise 26-11: Broadway Shows

a. The labeled scatterplot of *gross receipts* vs. *percentage capacity* follows:

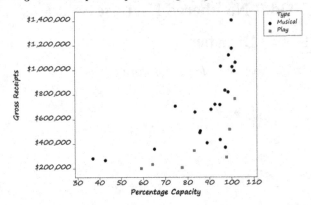

b. There is a moderate, positive, curved association between *gross receipts* and *percentage capacity*.

c. Almost without exception, among shows with similar percentage capacities, musicals tend to take in more money than do plays (although there are only seven plays in this list).

Exercise 26-13: Weighty Feelings

a. About right (*weight*) is represented by a diamond; overweight is represented by a circle; and underweight is represented by a square.
   Those respondents who feel underweight would be those who are below the weights of others at the same heights. Those who feel overweight should be heavier than others at the same heights, and those who feel about right will be in the middle.

b. The scatterplot displays a positive, linear association between *height* and *feelings about weight*. You also notice that very few respondents considered themselves underweight. There is a surprising consistency of opinion about the weights. There is a surprising consistency of opinion about the weights. At most heights, there are intervals of weights about which all respondents felt the same.

Exercise 26-15: Fast-Food Sandwiches

a. Many answers are possible. The possible labeled scatterplots are shown below. Students should display any *one* of these, along with a discussion of what the graph reveals.

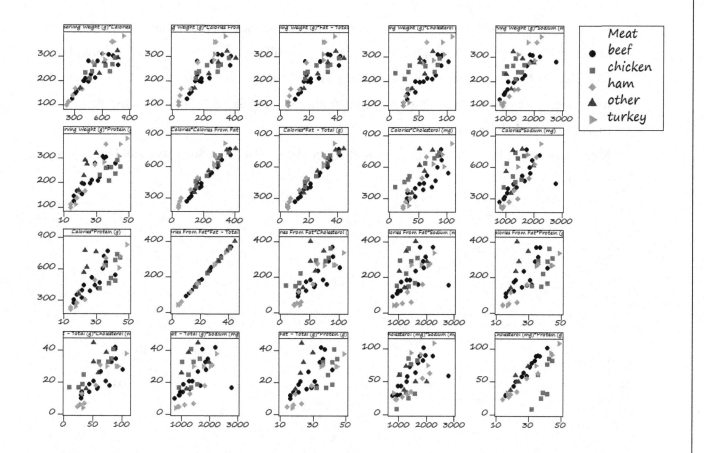

Exercise 26-17: Peanut Butter

a. The observational units are the brands of peanut butter.

b. The variables are classified as follows: *cost* (quantitative), *sodium* (quantitative), *quality* (quantitative), *crunchy/creamy* (binary categorical), *regular/natural* (binary categorical), *salted/unsalted* (binary categorical).

c. Many answers are possible. The possible scatterplots are shown here. Students should create one of

these and discuss what the graph reveals:

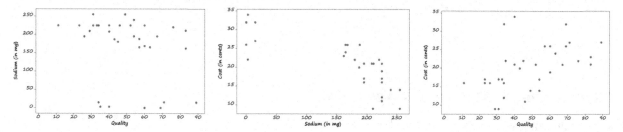

For example, the association between *cost* and *quality* is fairly strong, positive and linear. The association between *cost* and *sodium* is not as strong; overall the association appears negative, but there is a cluster of salt-free brands that should perhaps be investigated separately. This is especially true in examining the *sodium* and *quality* relationship. Without the low-sodium brands, the relationship could be negative.

d. Many answers are possible. The possible scatterplots are shown here. Students should create one of these and discuss it:

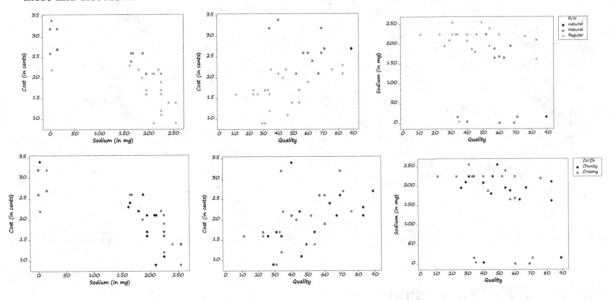

For example, these scatterplots reveal that the *chunky* vs. *creamy* brands don't appear to behave very differently.

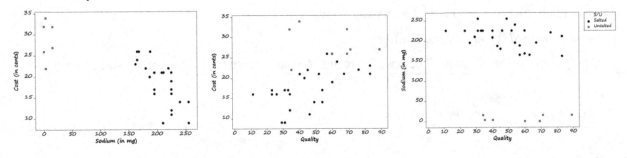

Exercise 26-19: Maternal Oxygenation

a. A scatterplot with the $y = x$ line is shown here:

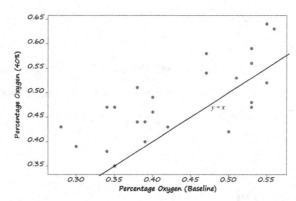

Because the majority of this data falls above the $y = x$ line, you know that fetuses tend to have a higher percentage of oxygen after the mother is administered 40% oxygen than at the baseline measurement.

b. The scatterplot does support the conclusion that fetuses with the lowest initial oxygen levels appeared to increase their oxygen percentages the most because these fetuses (with the lowest initial oxygen levels) tend to be the furthest above the $y = x$ line. The data with the larger initial oxygen levels are the only ones in which the oxygen levels decreased after the mother was administered 40% oxygen.

Exercise 26-21: Comparison Shopping

a. The observational units are the products available at both grocery stores. The two quantitative variables recorded are *Lucky's price* and *Vons price*.

b. The scatterplot of the *prices at Lucky's* vs. *prices at Vons* follows:

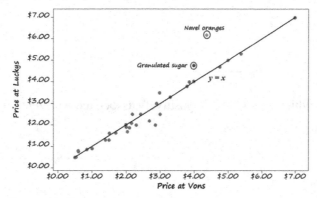

There is a very strong, positive, linear association between the prices of the products in these two stores. The prices of products are very similar, as indicated by so many of the pairs of prices being close to or on the $y = x$ line. It appears that the prices at Vons may be, on average, slightly less than those at Lucky's because there are more points below the $y = x$ line than above (excluding two outliers), but it is difficult to tell.

c. There are two suspicious products: navel oranges ($6.18, $4.36) and granulated sugar ($4.75, $3.99). These products are suspicious because their price at Lucky's is so much more than their price at Vons. Perhaps they were on sale at Vons that week, or perhaps a mistake was made in recording these prices.

Exercise 26-23: Age and Diet

Answers will vary. The following are a representative set:

a. A scatterplot in which *weight loss* is moderately positively associated with *age* is shown here:

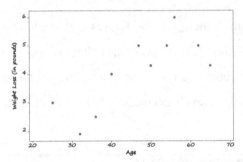

b. A scatterplot in which *weight loss* is not (linearly) associated with *age* is shown here:

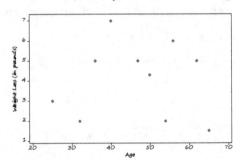

c. Here is a scatterplot in which *weight loss* is positively associated with *age* for women and negatively associated with *age* for men:

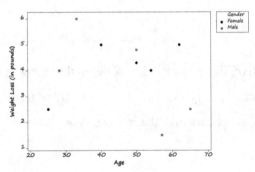

d. Here is a scatterplot in which *weight loss* is negatively associated with *age* for women and not associated with *age* for men:

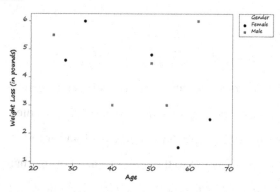

Exercise 26-25: Airport Traffic

a. The scatterplot shows a strong, positive, linear association between *number of passengers* and *year*.

b. The association is always roughly linear, but the particular linear relationship tends to change roughly every two decades. There is a linear relationship shown during the years 1941–1959, but there is a change to a different linear relationship with a steeper slope during the years 1960–1979. The slope of the linear relationship from 1980–2000 is much less steep than in the previous 40 years, and after 2000, the linear relationship changes again to a line with the steepest slope and a very different intercept.

c. The decrease in the number of passengers in the early 2000s is most likely due to the tragedies that occurred on September 11, 2001.

Exercise 26-27: Cat Jumping

a. Here is the labeled scatterplot:

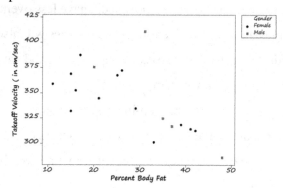

b. Here are comparative dotplots:

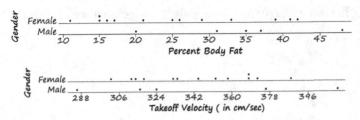

c. These dotplots indicate that the male cats tend to have a higher percentage of body fat than do the females; however there are only 5 male cats in this sample, whereas there are 13 female cats. The lower quartile for the males (25.5%) is greater than the median for the females (15.5%), and the female maximum (42%) is roughly equal to the upper quartile of the males (42.5%). The standard deviations for the males and females are very similar (10.13 % and 10.71%, respectively), but the mean for the male cats (34.2%) is substantially larger than that of the females (25.38%).

The variability in *takeoff velocities* is almost twice as large for the male cats as it is for the females ($s = 49.7$ cm/sec and 26.92 cm/sec, respectively), though the means are nearly identical (342.8 cm/sec and 343.46, respectively). The male cats have the smallest (286.3 cm/sec) and largest (410.8 cm/sec) *takeoff velocities* in this sample and 9 of the female cats have *takeoff velocities* within the interquartile range of the male cat velocities.

The scatterplot shows a moderately strong, negative, linear relationship between *takeoff velocity* and *percent body fat*. For 4 of the 5 male cats in this sample, this relationship is very strong, and the fifth cat is the only clear outlier in the collection of all cats. This male cat had the highest *takeoff velocity* (410.8 cm/sec), higher than both the male and female cats who had 10%-20% more body fat.

Exercise 26-29: Life Expectancy

a. *Life expectancy* vs. *fertility rate*: moderately strong negative association

Countries with high fertility rates tend to be less advanced, have less overall wealth, and do not tend to have access to sufficient health care. Citizens of such countries will also tend to have short life expectancies. Similarly, the "wealthier" countries with access to good national health care tend to have the longer life expectancies and lower fertility rates.

Life expectancy vs. *GDP*: moderate positive (non-linear) association

Countries with vibrant economies (large GDPs) tend to have access to top-notch health care and disaster relief, and also long life expectancies. The countries with less vibrant economies also tend to have the shortest life expectancies.

Life expectancy vs. *Internet users per 100 people*: moderate positive association

Countries with a large number of internet users tend to be the countries with vibrant economies (large GDP), and long life expectancies.

b. Answers will vary by student expectation, but in this case the strongest association is between *life expectancy* and *fertility rate*.

Exercise 26-31: Horse Prices

a. The oldest horse in this sample is female.

b. The most expensive horse in this sample is male.

c. The association between *price* and *age* for female horses is moderately strong, negative, and mostly linear.

d. The association between *price* and *age* for the male horses is moderately strong, positive and somewhat linear.

e. A 10-year-old male horse would probably cost more than a 10-year-old female horse. A typical 10-year-old male horse appears to cost in the ballpark of $30,000, whereas a typical 10-year-old female horse appears to cost around $15,000.

Exercise 26-33: Airline Maintenance

a. The observational units in the table are the airlines.

b. Here is the scatterplot:

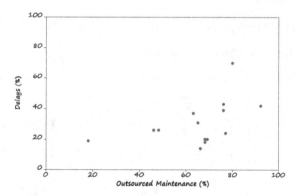

There is a weak to moderate, positive, linear association between the *percentage of delays due to the airline* and *percentage of maintenance outsourced*.

c. It is not reasonable to conclude there is a cause-and-effect relationship between these variables because this is an observational study and there are many potentially confounding variables that could explain the association.

Topic 27

Correlation Coefficient

Odd- Numbered Exercise Solutions

Exercise 27-9: Proximity to the Teacher

a. The correlation coefficient cannot be less than −1.0.

b. The order in which you state the variables will not change the correlation coefficient. The correlation coefficient of *x* and *y* is always the same as the correlation coefficient of *y* and *x*.

c. A correlation coefficient of −.8 indicates a *strong* negative association.

d. In order to compute a correlation coefficient, both variables must be quantitative.

e. If the correlation coefficient is −.8, then the association is negative, so students who sit farther away tend to score *lower*.

f. You can *never* conclude the existence of cause-and-effect relationship between two variables based solely on the value of the correlation coefficient. This can only be determined by knowing the type of study.

Exercise 27-11: Monopoly

a. The observational units are the properties on the Monopoly game board.

b. Five variables are listed for each observational unit.

c. Students should have one of the following scatterplots:

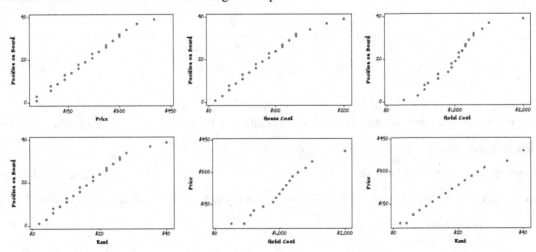

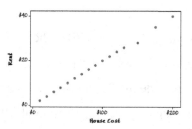 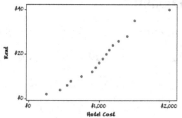 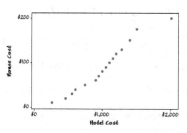

d. *position* and *price*: r = .995   *position* and *house*: r = .984   *position* and *hotel*: r = .964

   position and *rent*: r = .983   *price* and *hotel*: r = .978   *price* and *house*: r = .994

   rent and *house*: r = .999   *hotel* and *house*: r = .981   *price* and *rent*: r = .994

e. Answers will vary by student expectation. The correlation coefficient increases except in four cases. It is unchanged with *rent* /*house* and *rent*/ *price*, and it decreases for *rent*/*hotel* and *house*/*hotel*.

f. *Position* and *price*: r = .997   *Position* and *house:* r = .991   *position* and *hotel*: r = .985

   position and *rent*: r = .991   *price* and *hotel*: r = .984   *price* and *house*: r = .994

   rent and *house*: r = .999   *hotel* and *house*: r = .978   *price* and *rent*: r = .994

Exercise 27-13: Planetary Measurements

a. There is a strong, positive, curved association between planet *period of revolution* and *distance* from the sun.

b. A straight line would not be a reasonable summary of the relationship between *revolution* and *distance*. A curve would provide a much better fit.

c. The correlation coefficient of .989 does not mean a straight line is the best model for a reasonable summary of the relationship between these variables. You can see from the scatterplot that some curved model would provide a better fit. You should always look at a scatterplot, in conjunction with the correlation coefficient, to assess the form of the association.

Exercise 27-15: Broadway Shows

a. A: r = .721   B: r = .316   C: r = .947   D: r = .502   E: r = .407   F: r = .804

b. From smallest to largest correlation coefficient, you have B < E < D < A < F < C. This ordering does agree with the ordering based on the scatterplots alone in Exercise 26-10.

Exercise 27-17: *Challenger* Disaster

The correlation coefficient between *temperature* and *number of O-ring failures* is $-.561$. The correlation coefficient without the zero O-ring failure flights is $-.263$. The association between *temperature* and *number of O-ring failures* is weakened when you exclude the flights with no O-ring failures because those flights were all on days when the outdoor temperature was relatively high. The data are no longer present to show that warmer days tend to have no O-ring failures.

Exercise 27-19: Climatic Conditions

The completed table with correlation coefficients follows:

	Jan. High	Jan. Low	July High	July Low	Precip	Days Precip	Snow	Sun
Jan High	—	.965	.152	.554	−.073	−.572	−.807	.643
Jan. Low	—	—	−.072	.473	.002	−.460	−.825	.512
July High	—	—	—	.712	.114	−.130	−.080	.377
July Low	—	—	—	—	.243	−.345	−.613	.521
Precip	—	—	—	—	—	.695	−.157	−.506
Days Precip	—	—	—	—	—	—	.444	−.826
Snow	—	—	—	—	—	—	—	−.363

a. The strongest association is between *January high temperature* and *January low temperature* ($r = .965$).

b. The weakest association is between *January low temperature* and *annual precipitation* ($r = .002$).

c. The most useful variable for predicting *annual snowfall* would be *January low temperature* ($r = −.825$). The least useful variable would be *July high temperature* ($r = −.080$).

d. The most useful variable for predicting *July high temperature* would be *July low temperature* ($r = −.712$). The least useful variable would be *January low temperature* ($r = −.022$).

e. The scatterplot of *annual snowfall* vs. *annual precipitation* follows:

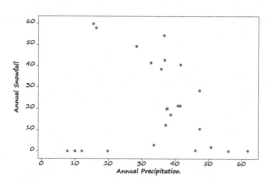

The correlation coefficient indicates a very weak positive relationship between these variables. A closer look at the scatterplot, however, indicates four western cities (Phoenix, San Diego, San Francisco, and Los Angeles) with no annual snowfall that might be artificially deflating the correlation coefficient. If you delete these cities and recomputed, you find a much stronger negative correlation coefficient of −.771.

Exercise 27-21: Digital Cameras

a. The scatterplot of *rating score* vs. *price* follows:

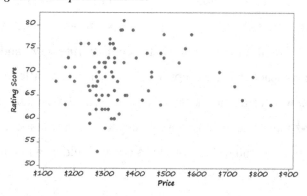

The correlation coefficient is .071, which indicates there is virtually no linear association between *rating score* and *price* of a digital camera. This makes sense when you see the scatterplot; a line would not fit this data well. A curve might fit moderately well.

b. The four scatterplots and correlation coefficients follow:

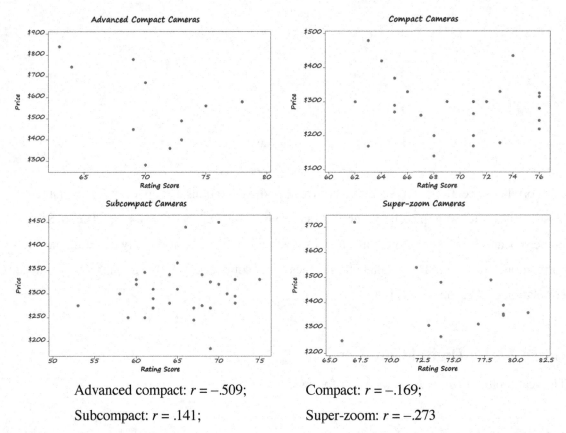

Advanced compact: $r = -.509$; Compact: $r = -.169$;

Subcompact: $r = .141$; Super-zoom: $r = -.273$

There is a moderate negative linear association between the *rating score* and *price* of the advanced compact cameras. The compact and subcompact cameras have very little linear association though subcompact cameras show a positive association whereas compact cameras show a negative association. The super-zoom cameras have a weak, negative, linear association according to the correlation coefficient, but the scatterplot indicates a curve would be a much better model for these data than a line.

Exercise 27-23: Emotional Intelligence

The correlation coefficients of .10 and .03 indicate there is virtually no linear association between *college GPA* and *emotional intelligence test scores* for either of these emotional intelligence tests; SAT scores are a much better predictor of college GPA than either of the emotional intelligence tests.

Exercise 27-25: House Prices

a. Many answers are possible. One point that could be added in order to increase the numerical value of the correlation coefficient is (2000 ft^2, $600,000). The additional of this point increases the correlation coefficient to .804. In order to increase the correlation coefficient, you need to add a point that follows the "line" formed by the previous points. The further you place that point out along the line, e.g. large size and large price, the larger the increase in the correlation coefficient.

b. Many answers are possible. One pair of points that could be added in order to decrease the correlation coefficient to a negative value is (450ft^2, $1,000,000) and (550 ft^2, $2,500,000). The addition of these two points makes the correlation coefficient equal to −.331. Both of the points have very small sizes, but extremely large prices, which go against the overall trend of the original scatterplot.

Exercise 27-27: Kentucky Derby

a. Here is a histogram of the winning speeds in miles per hour:

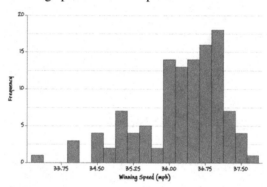

The distribution of wining speeds is skewed left with a minimum of 33.28 miles per hour and a maximum of 37.69 miles per hour. The mean is 26.22 mph and the standard deviation is .89 mph.

b. Here is the scatterplot:

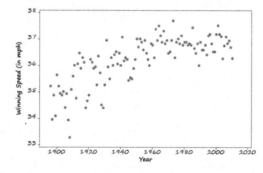

The correlation coefficient is .739. There is a fairly strong, curved, positive relationship between *winning speed* and *year*.

d. This scatterplot is identical to the scatterplot of *winning time* (in mph) vs. *year*, with the only change being the scale on the vertical axis. The correlation between *winning time* (in mph) and *year* is the same as the correlation between *winning time* (in mps) and *year*. This indicates a change in variable scale will have no effect on the correlation coefficient.

Exercise 27-29: In the News

a. Answers will vary.

Topic 28

Least Squares Regression

Odd- Numbered Exercise Solutions

Exercise 28-5: Textbook Prices

a. The regression line only predicts/estimates the price of textbooks; it will not, in general, provide the actual price of any particular textbook.

b. This statement does not consider the intercept in computing the predicted price of a textbook.

c. This statement does not consider the intercept in computing the predicted price of a textbook. The predicted price of a one-page textbook would be −$3.42 + $0.147 = −$1.95. This is an unusual prediction because you are extrapolating; one page is not a reasonable number of pages for any textbook (and is not similar to the *page* values observed in the dataset).

d. The last phrase of this sentence should be "meaning the predicted number of pages increases by $0.147 *for each additional page in the textbook.*"

e. This is a completely invalid interpretation of r^2. There is no measure of the percentage of textbook prices that are correctly predicted by the regression line (which would be the percentage of observations actually falling on the regression line). The coefficient of determination measures the variability in textbook prices explained by the linear relationship with the number of pages.

f. The coefficient of determination does not measure the percentage of points that fall close to the regression line.

g. All of the textbooks contain pages! The coefficient of determination does not measure the percentage of textbooks that contain pages.

h. There are some words missing in this statement. It should read: The coefficient of determination is r^2 = .667, meaning 67.7% of the *variability in* textbook prices in explained by the least squares line with number of pages.

·i. The word "prices" (the response variable) is missing in this statement. It should read: The correlation coefficient is r^2 = .667, meaning 67.7% of the variability in textbook *prices* is explained by the least squares line with number of pages.

Exercise 28-7: Airfares

a. For Jacksonville's fitted value, you calculate $187.80 + \$.054/\text{mi} \times (2604 \text{ mi}) = \$328.42$.

   For the residual, you calculate $288 - \$328.42 = -\$40.42$.

b. New Orleans has the largest (absolute) residual. Its residual value is $414 - \$300.23 = \$113.77$

c. The residual plot of *residual* vs. *distance* follows:

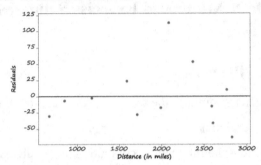

   The residuals do tend to increase in absolute value as the distance increases, but this may be due to the one extreme outlier (New Orleans).

Exercise 28-9: Airfares

a. If $500 is added to each airfare, the slope would not change because r, s_x and s_y would not change. The intercept would become

$$\$792.5 - \frac{\$.054}{\text{mi}} \times (2436 \text{ mi}) = \$667.76.$$

Therefore the regression equation would become

$$airfare = \$687.80 + \$.054/\text{mi} \times (distance).$$

b. If each airfare is doubled s_y, $\bar{x}$, and $\bar{y}$ would also double. This would result in both the slope and intercept coefficients being twice as large, so the regression equation would become

$$airfare = \$375.60 + \$.108/\text{mi} \times (distance).$$

c. If each distance were cut in half, s_x would be half as large, and therefore the slope coefficient would double. The intercept coefficient would not change because $\bar{x}$ would also be half as large, and therefore the product of b and $\bar{x}$ would be unchanged. The regression equation becomes:

$$airfare = \$187.80 + \$.108/\text{mi} \times (distance).$$

d. If 1000 miles was added to each distance, the slope would not change because r, s_x, and s_y would not change. The intercept would become

$$\$292.5 - \frac{\$.054}{\text{mi}} \times (1000 + 1936 \, \text{mi}) = \$133.73.$$

Therefore the regression equation would become

$$airfare = \$133.73 + \$.054/\text{mi} \times (distance).$$

Exercise 28-11: Electricity Bills

a. The following dotplot displays the electric bill charges:

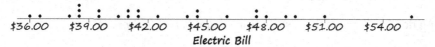

The distribution of electric bills has a slight skew to the right. The bills range from a minimum of about \$36 to a maximum of about \$55.5. The mean electric bill for these 28 months is \$43.18 and the standard deviation is \$.997 (just about a dollar).

b. The scatterplot of *electric bill* vs. *average temperature* follows:

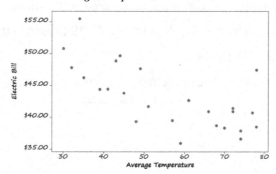

The scatterplot shows a moderate, negative, linear association between the *electric bill* and the *average temperature*.

c. Slope $b = -.695 \times \dfrac{4.99}{16.21} = -0.21395.$

Intercept $a = \$43.18 + 0.21395 \times (55.88) = \$55.14.$

Least squares regression equation: *city mpg* = $55.14 - 0.21395 \times$ (*avg. temp*)

d. The slope is -0.21395. This means that, on average, the predicted electric bill falls \$.21 for each rise of one degree in the average temperature.

e. The intercept indicates the predicted bill when the average temperature is zero degrees Fahrenheit. This does not tell us much because of the large extrapolation being used.

f. For the March 1992 fitted value, you calculate $55.14 – $0.21395/°F × (41°F) = $446.37. For the residual, you calculate $44.43 – $46.37 = $–1.94.

g. The month with the greatest fitted value would be the month with the lowest average temperature. According to the scatterplot, this is March 1993.

h. The percentage of variability in electric bills explained by the regression line with average temperature is $r^2 = (-.695)^2 \approx 48.3\%$.

Exercise 28-13: House Prices

a. The following dotplot displays the distribution of *number of bedrooms*:

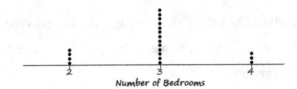

Number of Bedrooms

This is an almost perfectly symmetric distribution, centered at 3 bedrooms, with a minimum of 2 bedrooms and a maximum of 4 bedrooms. All but 6 of the 19 houses in this sample have 3 bedrooms. The standard deviation is 0.577 bedrooms.

b. The correlation coefficient is .499. The scatterplot of *house price* vs. *number of bedrooms* follows:

Scatterplot of Price vs Bedrooms

The scatterplot shows a positive association between number of bedrooms and house price, but because of the granularity in the number of bedrooms, a standard regression line is not likely to be a good model for this relationship. You also don't have very many houses with four bedrooms to confirm whether the average price for a four bedroom house is larger or smaller than for a house with three bedrooms.

c. The regression equation is *price* = 308,156 + 61,083 × (*number of bedrooms*); $r^2 = 24.9\%$.

d. The residual plot follows:

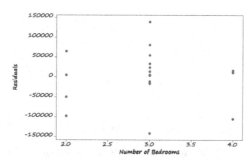

The residual plot indicates that a linear model is not appropriate here. The residuals for three bedroom houses are almost all positive or near zero, whereas the two and four bedroom houses tend to have large negative residuals.

e. The following dotplot displays the distribution of the *number of bathrooms*:

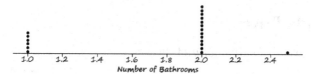

The distribution of bathrooms is skewed left with a minimum of 1 bathroom and a maximum of 2.5 bathrooms. Only one house had 2.5 baths, whereas 13 of the 19 houses had 2 bathrooms. The mean number of bathrooms per house is 1.76 and the standard deviation is .482 bathrooms.

The correlation coefficient is .711. The scatterplot of *house price* vs. *number of bathrooms* follows:

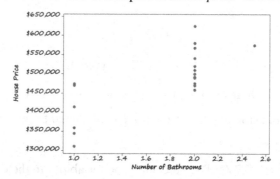

This scatterplot shows a much stronger positive linear association than did the *house price* vs. *number of bedrooms* plot. There is still a problem with the granularity in the number of bathrooms, and the single house with 2.5 bathrooms may be very influential.

The regression equation is $price = 307,427 + 104,347 \times (number\ of\ bathrooms)$; $r^2 = 50.6\%$.

The residual plot follows:

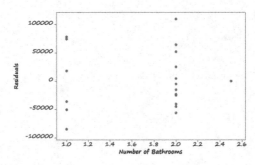

The residual plot does not show an obvious pattern to indicate the linear model is inappropriate, but due to the granularity of the data, it is difficult to tell much from the residual plot.

f. Based on the residual plots and values of r^2, the *house size* seems to be the best of these three explanatory variables for predicting *house price*.

Exercise 28-15: Honda Prices

a. The scatterplot *price* vs. *year of manufacture* follows:

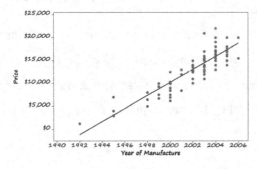

The regression equation is *price* = –2,840,147 + 1425 × (*year*).

The slope is $1425/year, which means the price is predicted to increase by $1425 on average for each additional year in the manufacture date. Or, the price is predicted to decrease by $1425 for each additional year in the car's age.

You calculate $r^2 = (.875)^2 \Rightarrow 76.7\%$. The percentage of variability in the car prices explained by the linear regression with year of manufacture is 76.6%.

Predicted price of a 1998 Honda Civic = $–2,840,147 + 1425 × (1998) = $7003.

Predicted price of a 2003 Honda Civic = $–2,840,147 + 1425 × (2003) = $14,128.

The outlier (in year) is the 1992 car (1992, $1200). With this car removed, the regression equation is

price = –2,911,782 + 1461 × (*year*).

The value of r^2 is 75.1%. Because the slope and coefficient of determination did not change much, you conclude this point is not very influential in the regression.

The residual plot follows:

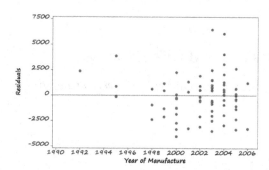

This residual plot does not show any strong patterns to indicate a linear regression is not the best model for describing this relationship.

Exercise 28-17: Box Office Blockbusters

The scatterplot of *overall gross income* vs. *opening weekend gross* follows:

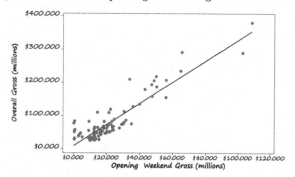

The scatterplot shows a moderately strong positive relationship between these variables for these blockbuster movies. The relationship does not appear completely linear, however, as there is a large cluster of movies with small opening values lying below the regression line.

The regression equation is *predicted overall gross* = 9513 + 3.182 × (*opening weekend gross*). The value of r^2 is 82.9%. A residual plot follows:

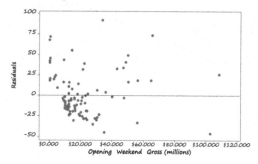

This residual plot indicates that a linear model may not be the best fit for this relationship. Movies that grossed less than $40 million in the opening weekend tend to have negative residuals, whereas movies that grossed more tend to have positive residuals.

Exercise 28-19: Televisions and Life Expectancy

a. The scatterplot of *life expectancy* vs. *number of televisions per 1000 people* follows:

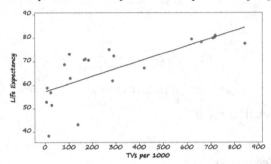

The regression equation is *predicted life expectancy* = 57.34 + 0.03244 × (*TVs per 1000*).

The value of r^2 is 55.2%. The residual plot follows:

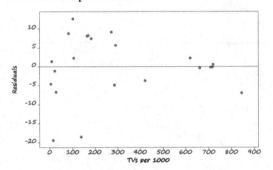

The least squares line does not provide a reasonable model of the relationship between these variables because the relationship is curved, as seen by the pattern of negative-positive-negative residuals.

b. A scatterplot of *life expectancy* vs. log(*TVs per 1000*) follows:

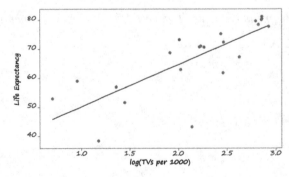

The regression equation is *predicted life expectancy* $= 35.79 + 14.46 \times \log_{10}(TVs\ per\ 1000)$. The value of r^2 is 60.5%. The residual plot follows:

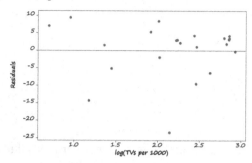

This residual plot does not reveal any definite patterns, so a linear model between *life expectancy* and log(*TVs per 1000*) seems appropriate.

c. The scatterplot of *life expectancy* vs. $\sqrt{TVs\ per\ 1000}$ follows:

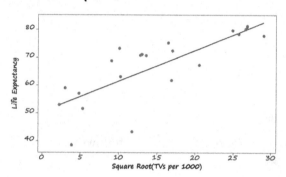

The regression equation is *predicted life expectancy* $50.32 + 1.111 \times \sqrt{TVs\ per\ 1000}$. The value of r^2 is 62.2%. The residual plot follows:

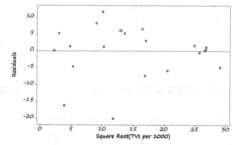

d. The least squares model is more appropriate for both of these transformed variables than with the original data because the transformed scatterplots are much more linear and the residual plots display fairly random scatter. The square root transformation is slightly better than the log transformation because it explains a slightly greater proportion of the variability in life expectancies.

Exercise 28-21: Planetary Measurement

a. The scatterplot of *period of revolution* vs. *distance* follows:

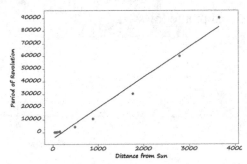

There is a very strong positive relationship between these variables, but it does not appear to be linear.

b. The scatterplot of *period of revolution* vs. *distance*² follows:

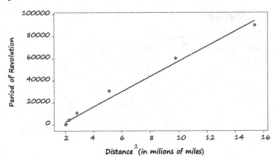

This relationship is *not* linear.

c. The scatterplot of *period of revolution* vs. *distance*^1.5 follows:

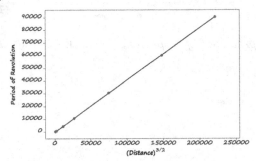

The power that appears to produce the most linear relationship is 1.5 or 3/2.

d. The regression equation is *revolution* = 11.9 + .4094 × (*distance*)^{1.5}; the value of r^2 is 100%.

e. The residual plot follows:

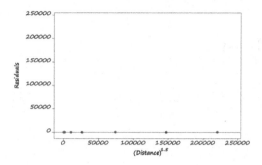

The residuals are all (relatively) very small. (Note: The scales on the vertical and horizontal axes in the residual plot were adjusted to be identical, in order to observe the relative size of the residuals.) Seven of the nine residuals are negative (including the first six), but because there are only nine data values, it's difficult to know whether this behavior indicates a pattern.

Exercise 28-23: Gestation and Longevity

a. For a human being, you calculate *gestation* = 54.53 + 10.34 × (75) = 830 days ≈ 2.27 years.

b. This is not a reasonable prediction because you are extrapolating and the linear relationship clearly does not hold for animals with such long average lifetimes. The longevities of the mammals in the given data range from about 1 to 40 years, and the average lifespan of a human being (75 years) is very far outside this range.

Exercise 28-25: Wrongful Conclusions

a. Because the mean is the sum of all residuals divided by the number of residuals, if the sum of the residuals is zero, the mean of the residuals must also be zero.

b. The median does not necessarily have to be zero in this case. For example, suppose the residuals are $\{-2, -2, -2, -2, -2, 10\}$. The sum of the residuals is zero, but the median residual value is -2, not 0.

c. The observation with the greatest value of the predictor variable will have the greatest fitted value only if the slope coefficient of the regression line is *positive*.

d. There is no way of predicting how large or small the residual of the observation with the greatest value of the response variable will be, relative to that of the other observations.

e. If the value of the predictor variable is the mean ($\bar{x}$), then the fitted value for that observation will equal the mean of the response variable ($\bar{y}$). However, you have no way of knowing the observed value of the response variable for that observation. It could be close to $\bar{y}$, or far from it. So it is possible for such an observation to have the greatest (absolute) residual value.

Exercise 28-27: Word Twist

a. Here is the scatterplot:

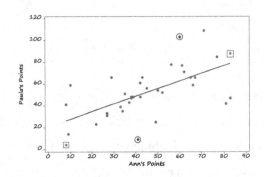

b. The least squares regression line equation is *Paula's score* = 20.78 + 0.7022 × (*Ann's score*).

c. The observation with the largest positive residual occurred in round 16. The observation with the largest (absolute) negative residual occurred in round 10. These points are circled in blue on the scatterplot.

d. Round 27 has the largest fitted value, and round 9 has the smallest fitted value. These points have a green square around them in the scatterplot. You know these points have the largest and smallest fitted values because there is a positive relationship between *Ann's score* and *Paula's score*, which means the fitted values will increase from smallest to largest as *Ann's score* increases from smallest to largest. (Ann's lowest score, 8, occurred in round 9, and her highest score, 83, occurred in round 27.)

Exercise 28-29: Cat Jumping

a. Here is the scatterplot:

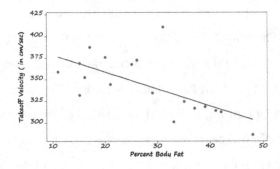

The least squares regression line equation is *takeoff velocity* = 397.7 − 1.953 × (*percent body fat*).

b. The slope coefficient is −1.953, which means the *takeoff velocity* is predicted to decrease by 1.953 cm/sec on average for each one percent increase in *body fat*.

c. The coefficient of determination is $r^2 = 42.3\%$. This means that 42.3% of the percentage of variability in the *takeoff velocities* is explained by this linear regression with *percent body fat*.

d. Here is the residual plot:

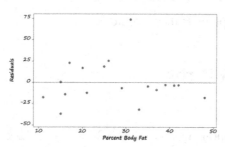

The residual plot indicates when the percentage of body fat is more than 32%, the residuals are all negative. However, there is a clear outlier (31%, 410.8 cm/sec) that is quite influential in this regression. If this cat is removed from the analysis, the regression equation becomes

$$takeoff\ velocity = 396.6 - 2.074 \times (percent\ body\ fat),$$

and r^2 jumps to 64.2%.

There is no clear pattern in the residual plot with the outlier removed:

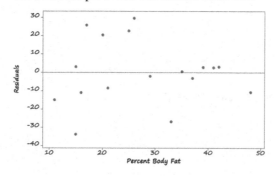

Exercise 28-31: Airport Traffic

a. Here is a rough sketch of the residual plot:

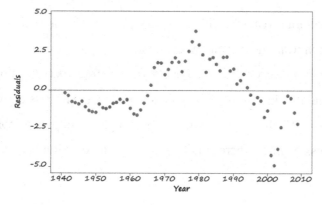

b. The years 2000–2010 tend to have the largest (absolute) negative residuals. This makes sense because these are the years when air traffic dropped off dramatically after September 11, 2001.

c. The years 1970 – 1990 all have positive residuals which means there was more air traffic during these years than the regression line predicts there would be.

Exercise 28-33: Kentucky Derby

a. Years before 1960:

Least squares regression equation: *winning time* = 342.3 – 0.1122 × (*year*); r^2 = 44.8%

Years 1960–2010:

Least squares regression equation: *winning time* = 98.43 + 0.01194 × (*year*); r^2 = 2.3%.

b. Here is the residual plot for the years before 1960:

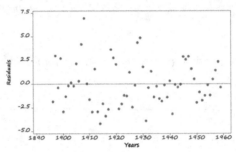

Here is the residual plot for the years 1960–2010:

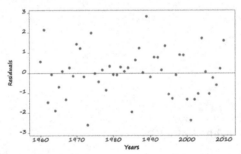

There is no obvious pattern revealed in either residual plot, so a linear model would be reasonable for the relationship between *winning time* and *year* in both cases.

c. The two regression lines are very different. For the years before 1960, the slope is negative, and almost half of the variability in *winning times* is explained by the regression with *years*. However, for the years from 1960 to 2010, the slope is *positive* and only 2.3% of the variability in *winning times* is explained by the regression line. There simply is no linear association between these variables from 1960–2000.

Exercise 28-35: Life Expectancy

For *Internet users per 100*:

Here is the scatterplot with a regression line superimposed:

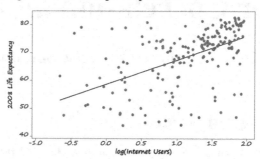

The regression equation is *predicted life expectancy* = 58.71 + 8.708 × log(*Internet users per 100*), and $r^2 = 27.4\%$.

This model does not provide a substantial improvement over the linear model using *Internet users per 100*. Although the curvature in the original scatterplot has been straightened, the coefficient of determination remains virtually unchanged (23.9% vs. 27.3%).

For *GDP*:

Here is the scatterplot with a regression line superimposed:

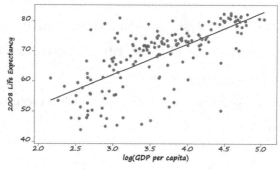

The regression equation is *predicted life expectancy* = 31.69 + 10.18 × log(*GDP*), and $r^2 = 47.5\%$.

This model is a substantial improvement over the model using *GDP*. The obvious curvature in the original scatterplot has been straightened out, and r^2 has almost doubled (from 26.6% to 47.5%).

Exercise 28-37: Own Your Own

Answers will vary.

Topic 29

Inference for Correlation and Regression

Odd- Numbered Exercise Solutions

Exercise 29-7: Studying and Grades

a. H_0: There is no correlation between *hours studied* and *GPA* in the population of all students at UOP. In symbols, $\rho = 0$.

   H_a: There is a correlation between *hours studied* and *GPA* in the population of all students at UOP. In symbols, $\rho \neq 0$.

   The test statistic is $t = \dfrac{-.343 \times \sqrt{78}}{\sqrt{1-(-.343)^2}} \approx \dfrac{3.029}{\sqrt{1-(.1176)}} = 3.22$.

   Using Table III with 60 degrees of freedom, $2 \times (.001) < p\text{-value} < 2 \times (.005)$, or $.002 < p\text{-value} < .01$.

b. This is the same value of the test statistic that you found in Activity 29-1. However, this *p*-value is twice as large as the *p*-values you found in Activity 29-1 because this is a two-sided test (notice, however, that the one-sided *p*-values are the same).

c. Answers will vary by student expectation, but students should expect the *p*-value to be larger because the sample size is smaller.

d. The test statistic is $t = \dfrac{-.343 \times \sqrt{18}}{\sqrt{1-(-.343)^2}} \approx \dfrac{1.46}{\sqrt{1-(.1176)}} = 1.55$.

   Using Table III with 18 degrees of freedom, $2 \times (.05) < p\text{-value} < 2 \times (.1)$, or $.1 < p\text{-value} < .2$. The *p*-value is indeed larger.

e. Answers will vary, but students should expect the *p*-value to be smaller, because the sample size is larger.

   The test statistic is $t = \dfrac{-.343 \times \sqrt{198}}{\sqrt{1-(-.343)^2}} \approx \dfrac{4.83}{\sqrt{1-(.1176)}} = 5.14$.

   Using Table III with 100 degrees of freedom, $p\text{-value} < 2 \times (.0005)$. The *p*-value is indeed smaller.

f. In order to be statistically significant at the $\alpha = .05$ level with $n = 5000$, performing a two-sided test, you need $t = 1.96$. So

$$1.96 \leq \frac{r \times \sqrt{4998}}{\sqrt{1 - r^2}}$$

which means

$$1.96^2 \left(1 - r^2\right) \leq r^2 \times 4998 \implies \frac{1.96^2}{4998 + 1.96^2} \leq r^2 \implies r \geq .027719$$

With such a large sample size, the correlation coefficient does not need to be very large in order to be "statistically significant."

Exercise 29-9: Draft Lottery

a. H_0: There was no correlation between *birth date* and *draft number* in the 1970 draft process. In symbols, $\rho = 0$.

H_a: There was a correlation between *birth date* and *draft number* in the 1970 draft process. In symbols, $\rho \neq 0$.

The test statistic is $t = \dfrac{-.226 \times \sqrt{364}}{\sqrt{1 - (-.226)^2}} \approx \dfrac{-4.31}{\sqrt{1 - (.051)}} = -4.43$.

Using Table III with 100 degrees of freedom, p-value $< 2 \times (.0005) = .001$.

With such a small p-value, reject H_0. You conclude there is a statistically significant correlation between *birth date* and *draft number* in the 1970 draft process (or that the observed sample correlation did not occur by random chance alone).

b. *Note*: There were 365 days in 1971.

H_0: There was no correlation between *birth date* and *draft number* in the 1971 draft process. In symbols, $\rho = 0$.

H_a: There was a correlation between *birth date* and *draft number* in the 1971 draft process. In symbols, $\rho \neq 0$.

The test statistic is $t = \dfrac{-.014 \times \sqrt{363}}{\sqrt{1 - (-.014)^2}} \approx \dfrac{.267}{\sqrt{1 - (.000196)}} = 0.27$.

Using Table III with 100 degrees of freedom, p-value $> 2 \times .2 = .4$.

As the p-value is not small, do not reject H_0. You conclude there is no evidence of a correlation between *birth date* and *draft number* in the 1971 draft process.

c. Based on these *p*-values, reject the null hypothesis with the 1970 draft process, but not with the 1971 draft lottery. There is strong statistical evidence the 1970 draft process was not fair, but no reason to doubt that the 1971 draft process was fair.

Exercise 29-11: Honda Prices

a. The regression equation is $price = -2,840,147 + 1425 \times (year)$.

   The slope coefficient is $+1425$, which means the predicted price of a used Honda Civic falls by an average of \$1425 for each additional year of age on the car.

b. A 95% CI for β is $b \pm t^*_{60} \times SE_b = 1425 \pm 2.000 \times (78.86) = (1267.28, 1582.72)$.

c. This interval includes only positive values. Based on this interval, you conclude zero is not a plausible value for the population slope coefficient (β), so you reject the null hypothesis in a two-sided test that the population slope coefficient is equal to 0 at the $\alpha = .05$ significance level.

d. Here is a histogram and a probability plot of the residuals, as well as a plot of *residual* vs. *year of manufacture*:

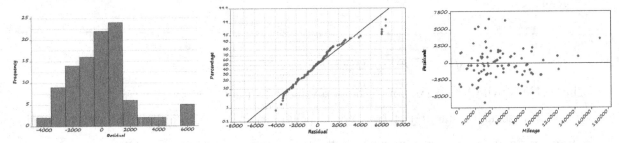

   Technical conditions: The residuals do not appear to be normally distributed, and the variability of the *y*'s does not appear to be quite the same across all years. So, technical conditions 3 and 4 are violated, although technical condition 2 is met (no curved pattern in the residuals).

e. It appears that *mileage* is a better explanatory variable for *price* than is *year of manufacture* because it comes closer to satisfying the technical conditions.

Exercise 29-13: Heights, Handspans, and Foot Lengths

Answers will vary by class. The following is one representative set.

a. The scatterplot of *height* vs. *foot length* follows:

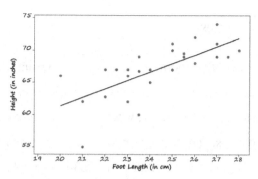

The scatterplot show a moderate positive linear association between *height* and *foot length* for these students.

b. The regression equation is *height* = 35.13 + 1.312 × (*foot length*).

c. The value of r^2 is 49.8%, which means that 49.8% of the variability in these *heights* is explained by the linear regression with *foot length*.

d. The residual plots follow:

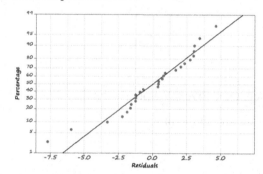
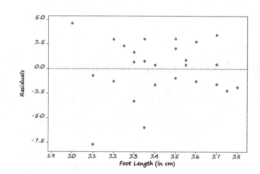

Technical conditions: The residual plot reveals no obvious pattern. The variability in the residuals appears to be similar across all *x*-values, and the residuals appear to be normally distributed. So, technical conditions 2–4 are satisfied. This was not a random sample of students from this school as it was one statistics class, but it might be reasonable to consider it a representative sample with respect to these variables.

e. These coefficients are statistics because they are calculated from a sample.

f. The null hypothesis is that there is no relationship between the *heights* and *foot lengths* for the population of all students at this school, or the slope of the population regression line between these two variables is 0. In symbols, $H_0: \beta = 0$.

The alternative hypothesis is that students at this school with longer feet tend to be taller, or the slope of the regression line between these two variables is positive. In symbols, $H_a: \beta > 0$.

The test statistic is $t = \dfrac{1.312}{0.2636} = 4.98$.

Using Table III with 25 degrees of freedom, p-value < .0005. Using Minitab, the p-value is .000. With the small p-value, reject H_0 at the $\alpha = .05$ level of significance. Conclude there is strong statistical evidence that the slope coefficient between *height* and *foot length* is positive in the population of all students at this school (assuming the sample was representative).

g. A 95% confidence interval for β is $1.312 \pm (2.06) \times (0.2636) = (0.769, 1.855)$. You are 95% confident the average increase in the *height* of students in this population is between .769 and 1.855 inches for each additional centimeter in *foot length*.

h. *Foot length* does seem to be a moderately useful predictor of *height* for students at this school, as long as this sample was representative. There is strong statistical evidence of a positive correlation between *height* and *foot length*, and you are 95% confident the average increase in *height* of a student in this population is between .769 and 1.855 inches for each additional centimeter of her or his *foot length*.

Exercise 29-15: Chip Melting

Answers will vary by class. The following is one representative example.

Here is a scatterplot with the regression line superimposed on the plot:

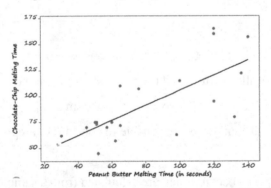

The regression equation is *predicted chocolate-chip melting time* = $33.23 + 0.7260 \times$ (*peanut-butter chip melting time*); $r^2 = 52.0\%$.

H_0: There is no correlation between the population melting times of *chocolate* and *peanut-butter* chips, or the slope of the population regression line between them is 0. In symbols; H_0: $\beta = 0$.

H_a: There is a positive correlation between these melting times. In symbols, H_a: $\beta > 0$.

The test statistic is $t = \dfrac{0.726}{0.1646} = 4.41$.

Using Table III with 18 degrees of freedom, the *p*-value < .0005. Using Minitab, the *p*-value is .000.

Test decision: The *p*-value is quite small, so reject H_0.

Conclusion in context: Conclude there is strong statistical evidence of a positive correlation between population melting times of these two types of chips.

Technical conditions:

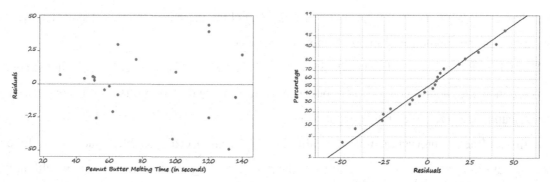

The residuals appear to be normally distributed, and there is no obvious pattern (curvature) in the residual plot, but the variability in the residuals is not very similar across all *x*-values. (The variation increases as the peanut-butter chip melting times increase.) This was not a random sample of students as it was one statistics class, but it could be a representative sample on these variables. Thus technical conditions 1–3 are satisfied, but not 4.

Exercise 29-17: Marriage Ages

a. The regression equation is *wife's age* = 2.446 + 0.8790 × (*husband's age*).

The value of r^2 is 89.1%.

A scatterplot of *wife's age* vs. *husband's age* follows:

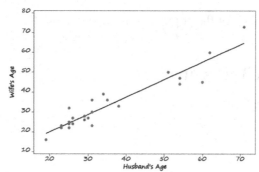

b. H_0: The slope of the regression line between *wife's age* and *husband's age* in the population is zero. In symbols, H_0: $\beta = 0$.

H_a: The slope of the regression line between these two variables is not 0. In symbols, H_a: $\beta \neq 0$.

The test statistic is $t = \dfrac{0.879}{0.06556} = 13.41$.

Using Table III with 22 degrees of freedom, p-value $< 2 \times .0005 = .001$. Using Minitab, the p-value is .000.

With this small p-value, reject the null hypothesis. Conclude there is extremely strong statistical evidence that the population slope coefficient is not zero.

c. A 95% CI for β is $0.879 \pm (2.074) \times (0.06556) = (0.743, 1.015)$. You are 95% confident the average increase in age for a wife is somewhere between 0.743 years and 1.105 years for each additional year her husband's age increases in this population.

d. Yes, the confidence interval includes the value 1. This means 1 is a plausible value for the slope of the population regression line. If the slope of the population regression line is 1, that would imply husbands who differ in age by one year would have wives who are predicted to differ in age by exactly one year (in the same direction) as well. In fact, husbands who differ in age by any number of years would have wives who are predicted to differ in age by the same number of years.

e. H_0: The slope of the regression line between *wife's age* and *husband's age* in the population is 1. In symbols, H_0: $\beta = 1$.

H_a: The slope of the regression line between these two variables is not 1. In symbols, H_a: $\beta \neq 1$.

The test statistic is $t = \dfrac{0.879 - 1}{0.06556} = -1.85$.

Using Table III with 22 degrees of freedom, $2 \times .025 < p$-value $< 2 \times .05 \Rightarrow .05 < p$-value < 10. Using Minitab, the p-value is .078.

Because the p-value $> .05$ (barely!), do not reject the null hypothesis at the $\alpha = .05$ significance level. Conclude 1 is a plausible value for the population slope coefficient at the 5% level.

Exercise 29-19: Cricket Thermometers

a. Here is a scatterplot of these data:

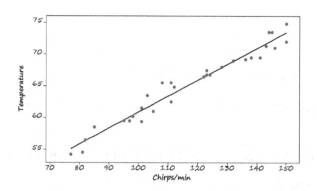

The regression equation is *temperature* = 35.78 + 0.2512 × (*chirps per minute*).

The value of r^2 is 95.7%.

H₀: The slope of the regression line between *temperature* and *chirps per minute* in the population is 0.

In symbols, H₀: β = 0.

Hₐ: There is a positive correlation between these two variables in the population.

In symbols, Hₐ: β > 0.

The test statistic is $t = \dfrac{0.25116}{0.01009} = 24.89$.

Using Table III with 27 degrees of freedom, *p*-value < .0005. Using Minitab, the *p*-value is .000. With this small *p*-value, reject the null hypothesis at any commonly used significance level. There is very strong statistical evidence of a positive correlation between *temperature* and *chirps per minute* in the population of all crickets.

The residual plots follow:

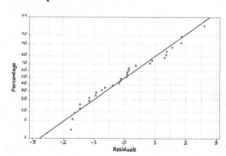 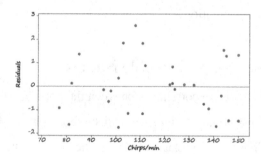

Technical conditions: The residuals appear to be normally distributed, there is no obvious curvature in the residual plot, and the residuals appear similar across all *x*-values. You are not told whether this was a random sample of crickets, but you can assume it is a representative sample with regard to these variables. Thus technical conditions 1–4 are all met.

b. A 99% CI for β is 0.25116 ± (2.771) × (0.01009) = (0.222, 0.280). You are 99% confident the average increase in *temperature* is somewhere between 0.222 °F and 0.280 °F for each additional *chirp per minute.*

Exercise 29-21: Utility Usage

a. H_0: The population slope between *average temperature* and *electricity usage* is zero. In symbols, H_0: $\beta = 0$.

H_a: The slope coefficient between these variables is negative. In symbols, H_a: $\beta < 0$.

The test statistic is $t = \dfrac{-0.12633}{0.03664} = -3.45$.

Using 80 degrees of freedom, the *p*-value is $< .0005$. Using Minitab, the *p*-value $= .001$.

Because the *p*-value $< .05$, reject the null hypothesis at the $\alpha = .05$ significance level. You have very strong statistical evidence of a negative correlation between *average temperature* and *electricity usage* for this population.

b. H_0: There is no correlation between *average temperature* and *gas usage* in the population. In symbols, H_0: $\beta = 0$.

H_a: The slope coefficient between these variables is negative. In symbols, H_a: $\beta < 0$.

The test statistic is $t = \dfrac{-0.21696}{0.01036} = -20.94$.

Using 80 degrees of freedom, the *p*-value $< .0005$. Using Minitab, the *p*-value $= .000$.

Because the *p*-value is so small, reject the null hypothesis at the $\alpha = .05$ significance level. You have very strong statistical evidence of a negative correlation between *average temperature* and *gas usage* for this population.

Exercise 29-23: Scrabble Names

a. H_0: There is no correlation between the population *number of letters in a name* and the number of *Scrabble points in that name*. In symbols, H_0: $\rho = 0$.

H_a: There is a correlation between the population *number of letters in a name* and the *number of Scrabble points in that name*. In symbols, H_a: $\rho \neq 0$.

b. The test statistic is $t = \dfrac{(.476) \cdot \sqrt{14}}{\sqrt{1 - (.476)^2}} = 2.03$.

c. Using Table III with 14 degrees of freedom, the $1.76 < p\text{-value} < 2.145 \Rightarrow 2 \times .025 < p\text{-value} < 2 \times .05$, so $.05 < p\text{-value} < .1$.

Because the *p*-value is greater than .05, do not reject the null hypothesis at the .05 significance level. You do not have sufficient evidence to conclude there is a correlation between the *number of letters in a name* and the *number of Scrabble points in that name* for the population.

d. If the alternative hypothesis were H_a: $\rho > 0$, the test statistic value would be the same (2.03), but the *p*-value would be half as large: $.025 < p\text{-value} < .05$. In this case, you would reject the null hypothesis at the $\alpha = .05$ significance level, and conclude there is a correlation between these two variables in the population.

e. Failing to reject the null hypothesis indicates the data provide no evidence of a *linear* relationship between *number of letters* and *Scrabble points* in students' names. There may be some other sort of relationship between these variables that would not be measured by the correlation coefficient.

Exercise 29-25: Own Your Own

Answers will vary.

Odd-Numbered Solutions to Additional Topics Exercises

Topic 30 Solutions

Homework Exercises

Exercise 30-7: Commuting to School

a. $\Pr(R_1) = .5$, $\Pr(R_2) = .4$, $\Pr(R_1$ or $R_2) = .6$

b.   Complement of {at least one of the lights will be red} = {not R_1 and not R_2}

   $\Pr(not\ R_1$ and $not\ R_2) = 1 - \Pr(R_1$ or $R_2) = 1 - .6 = .40$

c.   From parts a and b you have the following table:

	R_1	$Not\ R_1$	Total
R_2			.5
$Not\ R_2$		.40	
Total	.4		1.0

Here is the completed table:

	R_1	$Not\ R_1$	Total
R_2	.30	.20	.50
$Not\ R_2$	.10	.40	.50
Total	.40	.60	1.0

d. $.20 + .10 = .30$

e. $\Pr(R_2|R_1) = \Pr(R_2$ and $R_1)/\Pr(R_1) = .30/.40 = .75$

f.   The lights do not function independently because $.75 = \Pr(R_2|R_1) \neq \Pr(R_1) = .50$.

Exercise 30-9: Women Senators

a.   There are 12 women Democrats, so $\Pr(W$ and $D) = 12/100$.

   $\Pr(W$ or $D) = \Pr(W) + \Pr(D) - \Pr(W$ and $D) = 17/100 + 51/100 - 12/100 = 56/100 = .56$

   You can also construct a probability table:

	Republican (*R*)	Democrat (*D*)	Independent (*I*)	Total
Woman (*W*)	.05	.12	0	.17
Man (*M*)	.42	.39	.02	.83
Total	.47	.51	.02	1.00

b. $\Pr(D|W) = 12/17 \approx .706$

$\Pr(D) = 51/100 = .51$

If the senator is a woman, this increases the probability that the senator is a Democrat.

c.

$\Pr(R|M) = \Pr(R \text{ and } M)/\Pr(M) = .42/.83 \approx .506$

$\Pr(R) = .47$

So the probability of a male senator being a Republican is larger than the probability of a randomly selected senator being Republican.

d.   No, because $\Pr(D) \neq \Pr(D|W)$. Knowing the gender of the selected senator changes the probability you assign to their political party. Women senators are more likely to be Democrats and men senators are more likely to be Republicans.

Exercise 30-11: California Demographics

a.   Here is the completed table:

	Female	Male	Total
Under Age 18	.133	.140	.273
Age 18–64	.502 – .061– .133 = .308	.313	1 – .273 – .106 = .621
Age 65 or Older	.061	.045	.106
Total	.502	.498	1.000

b. $\Pr(female \text{ or } < 18 \text{ or } > 65) = .133 + .140 + .308 + .061 + .045 = .687$

c.   This is the complement of {male between the ages 18–64}, which has probability .313 from the table.

d. $\Pr(female) = .502$, $\Pr(age\ 65\ or\ older) = .106$

Pr(*female and age 65 or older*) = .045 ≠ .502 × .106 = .053, so the events are not quite

independent.

e.   These events are not mutually exclusive because there are females in this age category: Pr(*female*

and *age 18–65*) ≠ 0.

Exercise 30-13: Weighty Feelings

a.   Pr(*underweight | male*) = # *underweight males/*(*# males*) = 274/2855 ≈ .096

b.   Pr(*male | underweight*) = # *underweight males/*(*# underweight*) = 274/390 ≈ .702

c.   These are not mutually exclusive events because there are males that feel underweight. (To be

mutually exclusive, the table needs a zero in the (*male, feels underweight*) cell.)

d.   From part a, Pr(*underweight | male*) = .096, whereas Pr(*underweight*) = 390/5876 ≈ .066. So these

events are not quite independent.

Exercise 30-15: Daily Lottery

a.   Pr(*lose*) = 999/1000 = .999

Using the multiplication rule for independent events, Pr(*lose all 7 days*) = $.999^7$ ≈ .993.

b.   By the complement rule, Pr(*win at least once*) ≈ 1 − .993 = .007

c.   Pr(*win at least once*) = 1 − Pr(*lose all 30 days*) = $1 − .999^{30}$ ≈ .03

d.   Pr(*win at least once*) = 1 − Pr(*lose all 365 days*) = $1 − .999^{365}$ ≈ .306

e.   You want to find the value k so that $1 − .999^k > .5$. You can use trial and error to find that 693 is

the first value where this calculation exceeds .5. So you need to play 693 days for your

probability of winning at least once to exceed .5.

f. Similarly, you need to play 2302 days for the probability to exceed .9.

Exercise 30-17: Sports Series

a. Pr(*win* and *win*) = .7 × .7 = .49, because the results are independent from game to game.

b. Pr(*lose* and *win* and *win* OR *win* and *lose* and *win*) = (.3 × .7 × .7) + (.7 × .3 × .7) = .294, because these two outcomes are mutually exclusive.

c. Pr(*win in 2 games* or *win in 3 games*) = .49 + .294 = .784 by the addition rule, because these outcomes are mutually exclusive.

d. Pr(*Cache Cows win in 2 games* or *Domestic Shorthairs win in 2 games*) = (.3 × .3) + (.7 × .7) = .58

Exercise 30-19: Family Births

a. Pr(*second boy* | *first boy*) = Pr(*second boy*) = .5, because you are assuming independence from child to child.

b. Pr(*second boy* and *first boy*) = Pr(*second boy*) × Pr(*first boy*) = .5 × .5 = .25 (independence)

c. Pr(*first boy* or *second boy*) = Pr(*first boy*) + Pr(*second boy*) − Pr(*both boys*) = .5 + .5 − .25 = .75

d. Pr(*both boys* | *at least one boy*) = Pr(*both boys* and *at least one boy*)/Pr(*at least one boy*) = Pr(*both boys*)/Pr(*at least one boy*) = .25/.75 ≈ .333, because both boys is the only outcome also in {at least one boy}.

Exercise 30-21: Tennis Serves

a. You are given the following information:

Pr(*won point* | *first serve in*) = .76

Pr(*won point* | *first serve not in*) = .43

Pr(*first serve in*) = .64

You want to find Pr(*won point*). There are two ways he can win a point: if his first serve lands in or his first serve does not land in.

You have Pr(*won point*) = Pr(*won point* and *first serve in*) + Pr(*won point* and *first serve out*). You can add these probabilities because these are mutually exclusive events. Now you can use the multiplication rule to find these intersection probabilities:

Pr(*won point* | *first serve in*)Pr(*first serve in*) + Pr(*won point* | *first serve out*) Pr(*first serve out*)

= .76(.64) + .43(1 − .64) = .6412 (by the complement rule as well).

This indicates that he won 64.12% of the points he served. A probability table also shows these results:

	First Serve In	First Serve Out	Total
Won Point	.76(.64) = .4864	.43(.36) = .1548	.6412
Did Not Win Point	.64 − .4864 = .1536	.36 − .1548 = .2052	.3588
Total	.64	.36	1.0

b. You have Pr(*VW won point* | *VW served*) = .58, Pr(*VW won point* | *LD served*) = .47, and Pr(*VW served*) = .54.

Pr(*VW won point*) = Pr(*VW won point* | *VW served*)Pr(*VW served*) + Pr(*VW won point* | *LD served*)Pr(*LD served*) = .58(.54) + .47(1 − .54) = .5294

Odd-Numbered Solutions to Additional Topics Exercises

Topic 31 Solutions

Homework Exercises

Exercise 31-5: Lottery Games

a. Here is the probability distribution:

Possible Profits	499	−1
Probability	.001	.999

The expected value is $499 \times .001 + (-1) \times .999 = -\$0.50$. The *expected* outcome is to lose 50 cents each time you play this game.

b. If you play this game thousands and thousands of times, the long-run average profit will approach a loss of 50 cents per game.

c. $30 \times -.50 = -\$15$

d. $365 \times -.50 = -\$182.50$

Exercise 31-7: Rolling Dice

a–b. Here is the probability distribution and expected value for each variable:

For *S*:

Rolls	1,1	1,2; 2,1	1,3; 3,1; 2,2	1,4; 4,1; 2,3; 3,2	1,5; 5,1; 2,4; 4,2; 3,3	1,6; 6,1; 5,2; 2,5; 3,4; 4,3	2,6; 6,2; 3,5; 5,3; 4,4	3,6; 6,3; 4,5; 5,4	4,6; 6,4; 5,5	6,5; 5,6	6,6
Possible Outcomes	2	3	4	5	6	7	8	9	10	11	12
Probability	1/36	2/36	3/36	4/36	5/36	6/36	5/36	4/36	3/36	2/36	1/36

The expected value is $2 \times 1/36 + \ldots + 12 \times 1/36 = 7$. The long-run average value of the sum, over many, many rolls, will approach 7.

For *M*:

Rolls	1,1	1,2; 2,1; 2,2	1,3; 3;1, 2,3; 3,2; 3,3	1,4; 4,1; 2,4; 4,2; 3,4; 4,3; 4,4	1,5; 5,1 2,5; 5,2; 3,5; 5,3; 4,5; 5,4; 5,5	1,6; 6,1; 2,6; 6,2; 3,6; 6,3; 4,6; 6,4; 5,6; 6,5; 6,6
Possible Outcomes	1	2	3	4	5	6
Probability	1/36	3/36	5/36	7/36	9/36	11/36

The expected value is $1 \times 1/36 + \ldots 6 \times 11/36 = 161/36 \approx 4.47$. The long-run average value of the max, over many, many rolls, will approach about 4.47.

For *D*:

Rolls	1,1; 2,2;, ..., 6,6	1,2; 2,3; 3,4; 4,5; 5,6	1,3; 2,4; 3,5; 4,6	1,4; 2,5; 3,6	1,5; 2,6	1,6
Possible Outcomes	0	1	2	3	4	5
Probability	6/36	10/36	8/36	6/36	4/36	2/36

The expected value is $0 \times 6/36 + \ldots + 5 \times 2/36 = 70/36 \approx 1.94$. The long-run average value of the difference (in absolute value), over many, many rolls, will approach about 1.94.

c. The sum has the largest expected value. This makes sense because the sum totals both values rather than taking the max of one value or the difference between the values.

Exercise 31-9: Dice Rolling

a. Here is the probability distribution of *D*:

Rolls	1,1; ...; 6,6	all the rest
Possible Outcomes	1	0
Probability	6/36	30/36

b. The expected value is $1 \times 6/36 + 0 \times 30/36 = 6/36 \approx 0.167$. If you repeatedly roll the dice, the long-run average value for *I* will approach about 0.167.

c. Let *P* represent the profit. The expected value of *P* equals (*amount won if doubles*) $\times$ 6/36 + (*amount lost if not doubles*) $\times$ 30/36, where the amount lost is just the $1 bet. So you need *amount won* × 6/36 = (1) × 30/36. If you win $5 when the dice land on "doubles," the expected value of P equals 0.

Exercise 31-11: Basketball Shooting

a. The expected value of the number of points when attempting a two-point field goal is $2 \times .48 + 0 \times .52 = 0.96$. The expected value of the number of points when attempting a three-point field goal is $3 \times .34 + 0 \times .66 = 1.02$. The expected value of the number of points is higher for the three-point field goal.

b. The variance of two-point field goals is $(2 - .96)^2(.48) + (0 - .96)^2(.52) = 0.9984$.

standard deviation ≈ 0.999

The variance of three-point field goals is $(3 - 1.02)^2(.34) + (0 - 1.02)^2(.66) = 2.0196$.

standard deviation ≈ 1.42

c. In the long-run, this player will average more points with the three-point shot but also has more variability in the points scored.

d. You want $2 \times p$ to exceed 1.02, so the two-point shot is more advantageous if she makes at least 51% of those shots (assuming she continues to make 34% of the three-point shots).

Exercise 31-13: Selling Magazines

a–b. Here is the probability distribution for 2 magazines:

Magazines Sold	1	2
Profit	$0 + $.75 = $.75	$4
Probability	.1	.9

The expected profit is $0.75(.1) + 4(.9) = \$3.67$.

Here is the probability distribution for 3 magazines:

Magazines Sold	1	2	3
Profit	$1 \times \$4 - 3 \times \$2 + 2 \times \$.75 = -\$.5$	$2 \times \$4 - 3 \times \$2 + \$.75 = \$2.75$	$3 \times \$4 - 3 \times \$2 = \$6$
Probability	.1	.2	$1 - .1 - .2 = .7$

The expected profit is $-.5 \times .1 + 2.75 \times .2 + 6 \times .7 = \$4.70$.

Here is the probability distribution for 4 magazines:

Magazines Sold	1	2	3	4
Profit	$1 \times \$4 - 4 \times \$2 + 3 \times \$.75 = -\$1.75$	$2 \times \$4 - 4 \times \$2 + 2 \times \$.75 = \$1.50$	$3 \times \$4 - 4 \times \$2 + \$.75 = \$4.75$	$4 \times \$4 - 4 \times \$2 = \$8$
Probability	.1	.2	.3	.4

The expected profit is $-1.75(.1) + 1.5(.2) + 4.75(.3) + 8(.4) = \$4.75$.

Here is the probability distribution for 5 magazines:

Magazines Sold	1	2	3	4	5
Profit	$1 \times \$4 - 5 \times \$2 + 4 \times \$.75 = -\$3$	$2 \times \$4 - 5 \times \$2 + 3 \times \$.75 = \$.25$	$3 \times \$4 - 5 \times \$2 + 2 \times \$.75 = \$3.50$	$4 \times \$4 - 5 \times \$2 + \$.75 = \$6.75$	$5 \times \$4 - 5 \times \$2 = \$10$
Probability	.1	.2	.3	.3	.1

The expected profit is $-3(.1) + 25(.2) + 3.5(.3) + 6.75(.3) + 10(.1) = \$3.825$.

c. Now you should order 4 magazines, more than before.

d. The expected profit of $4.75 is larger than before. It makes sense for these values to be larger

    because there is a smaller penalty for not selling all of the magazines you order.

Exercise 31-15: Composite Testing

a. Here is the probability distribution:

Total Number of Tests Needed	1	21
Probability	$.9^{20} \approx .1216$	.8784

The expected value is $1(.1216) + 21(.8784) \approx 18.57$, which is still smaller than 20 tests.

b.	The expected value is $1 \times .9^n + (n+1)(1-.9^n) = .9^n + 1 - .9^n + n - n \times 9^n = 1 + n - n(.9^n)$, which you want to be greater than n.

$$1 + n - n(.9^n) > n$$

$$1 > n(.9^n)$$

This inequality is true for $n \geq 34$. Checking this result, $.9^{34} = .0278$. So the expected value becomes $.0278 + 35 \times (1 - .0278) = 34.11$, which exceeds 34.

c.	The expected value is $1 \times .95^{50} + 51 \times (1 - .95^{50}) \approx 47.15 < 50$, so that composite testing does reduce the expected number of tests.

d.	The expected value is $1 \times (1-p)^{50} + 51 \times (1 - (1-p)^{50}) = (1-p)^{50} + 51 - 51(1-p)^{50} = 51 - 50(1-p)^{50}$, which you want to be greater than 50 (and you know you need a value between .05 and .10). So you need $51 - 50(1-p)^{50} > 50$. This value for p needs to be greater than approximately .0753.

Exercise 31-17: Psychic Predictions

a.	Here is the probability distribution:

Outcomes	stock increases	stock decreases
Profit	20	0
Probability	.50	.50

The expected profit is $20(.5) + 0(.5) = \$10$.

b.	The expected profit for 1000 customers is $1000 \times \$10 = \$10,000$.

c.	The expected profit is $20(.20) + 0(.80) = \$4$. For 1000 customers, the expected profit is $\$4,000$.

Exercise 31-19: Family Births

a.	Here is the probability distribution:

Outcomes	GG	GB, BG	BB
Number of Girls	2	1	0
Probability	$.5^2 = .25$	$.5^2 + .5^2 = .50$	$.5^2 = .25$

b. The expected value is $2(.25) + 1(.50) + 0(.25) = 1$. If you were to repeatedly observe two-child families, the long-run average number of girls per family would approach 1.

c. The variance is $(2-1)^2(.25) + (1-1)^2(.50) + (0-1)^2(.25) = 0.5$.

standard deviation ≈ 0.707 girls

d. Here is the probability distribution:

Outcomes	GGGG	GGGB, ...	GGBB, ...	GBBB, ...	BBB
Number of Girls	4	3	2	1	0
Probability	$.5^4 = .0625$	$4 \times (.5)^3(.5) = .25$	$6 \times (.5)^2(.5)^2 = .375$	$4 \times (.5)(.5^3) = .25$	$.5^4 = .0625$

The expected value is $4(.0625) + 3(.25) + 2(.375) + 1(.25) + 0(.0625) = 2$ girls. If you were to repeatedly observe four-child families, the long-run average number of girls per family would approach 2.

The variance is $(4-2)^2(.0625) + (3-2)^2(.25) + (2-2)^2(.375) + (1-2)^2(.25) + (0-2)^2(.0625) = 1$.

standard deviation $= 1$ girl

e. The expected value is twice as large for the four-child family because you are observing twice as many children.

f. The variance is larger with the larger family. This makes sense because of the larger range of possible values for the number of girls.

Odd-Numbered Solutions to Additional Topics Exercises

Topic 32 Solutions

Homework Exercises

Exercise 32-7: Distinguishing Between Colas

a. $\Pr(X \geq 5)$, where X follows a binomial distribution with $n = 9$ and $\pi = 1/3$. Using technology, $\Pr(X \geq 5) \approx .145$.

b. $\Pr(X \geq 11)$, where X follows a binomial distribution with $n = 21$ and $\pi = 1/3$. Using technology, $\Pr(X \geq 11) \approx .056$.

c. $\Pr(X \geq 50)$, where X follows a binomial distribution with $n = 50$ and $\pi = 1/3$. Using technology, $\Pr(X \geq 50) \approx .0003$.

d. The probability decreases fairly dramatically as the sample size increases. With a larger number of trials, it is less likely that you will get lucky and perform so high above your expected correct identification probability.

Exercise 32-9: Distinguishing Between Colas

a. With 21 subjects, the probability of 12 or more subjects identifying the correct soda is less than .05; $\Pr(X \geq 12) \approx .0212$, but $\Pr(X \geq 11) \approx .0557$.

b. With 21 subjects, the probability of 13 or more subjects identifying the correct soda is less than .01; $\Pr(X \geq 13) \approx .0068$.

Exercise 32-11: Rolling Dice

a. Let X represent the number of rolls that result in a 6. You want to find $\Pr(X \geq 1)$.

b. X follows a binomial distribution with $n = 4$ and $\pi = 1/6$.

c. Pr(*at least one six*) ≠ Pr(*six on first roll*) + Pr(*six on first roll*) + Pr(*six on first roll*) + Pr(*six on first roll*) because these outcomes are not mutually exclusive. This would be an incorrect application of the addition rule. (*Note:* You also should not report 4/6 arising from $n\pi$ as this gives the expected value, not a probability.)

d. Using the binomial distribution and technology, $\Pr(X \geq 1) = 1 - \Pr(X = 0) \approx 1 - .4823 = .5177$.

e. Let Y represent the number of rolls (of both dice) that result in (6,6). You want to find $\Pr(X \geq 1)$, where X follows a binomial distribution with $n = 24$ and $\pi = 1/36$. You calculate $\Pr(X \geq 1) = 1 - \Pr(X = 0) = 1 - .5086 = .4914$.

f. These probabilities are similar, but the first bet is better in terms of having a higher probability of occurring.

Exercise 32-13: Free Candy Bars

a. $1000/(6000 - 18) \approx .167$

b. This probability is very close to $1/6 \approx .16667$.

c. $100/(600 - 18) \approx .172$

d. $10/(60 - 18) \approx .238$

Exercise 32-15: Heart Transplant Mortality

a. X follows a binomial distribution with $n = 10$ and $\pi = .85$.

b. Here is a graph of this probability distribution:

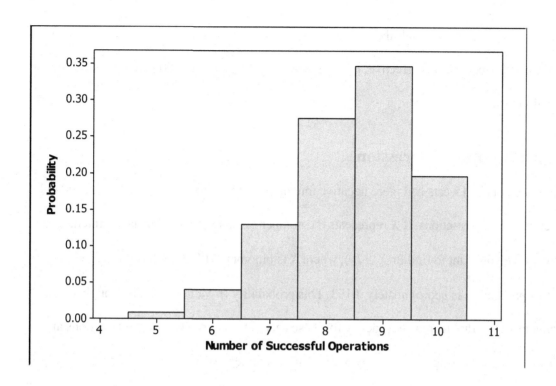

c. The most likely value of *X* is 9, with a probability of about .3474.

d. $\Pr(X \le 2) \approx .000000 + .000000 + .000008 \approx .000008$

e. This is an extremely small probability and provides very strong evidence against the claim that π = .85. It is *very* unlikely to find so few successful operations out of ten in a hospital where the survival rate is .85.

f. If *X* follows a binomial distribution with *n* = 371 and π = .85, $\Pr(X \le 292) \approx .0007389$.

g. It is very difficult for this hospital to say "we just got unlucky." This small probability gives very strong evidence against the claim that π = .85. Instead, you believe the survival rate is lower at this hospital compared to the national rate.

Exercise 32-17: Predicting Elections

The *p*-value is $\Pr(X \ge 189$, *where X is binomial with n = 279 and* $\pi = .5) < .0001$. This small *p*-value gives very strong evidence that the "competent face" method correctly predicts the winner in the U.S.

House of Representative more than half the time. It would be very surprising to have predicted the winner at least this often if the method was effective. (You are assuming the races in 2004 are representative of the larger election process.)

Exercise 32-19: Catnip Aggression

After the catnip, ten out of 15 cats had more negative interactions, three had the same amount, and only two had fewer negative interactions. If X represents the number of cats with more negative interactions after the catnip, then you want to find $\Pr(X \geq 10)$, where X is binomial with $n = 12$ and $\pi = .5$. Using technology, this probability is approximately .0193. This probability (p-value) is small enough to give you convincing evidence that there is a tendency for these cats to have more negative interactions after the catnip. It would be surprising for this many to have had more negative interactions after the catnip if there really was no such tendency.

Exercise 32-21: Alarming Wake-Up

a. In this study, 15 children escaped within five minutes for exactly one of the alarms, 13 of them to the mother's voice.

b. $\Pr(X \geq 13) \approx .0037$, where X follows a binomial distribution with $n = 15$ and $\pi = .5$.

c. This is a small probability and provides convincing evidence that the mother's voice is a more effective alarm. If the alarms were equally effective, it would be very surprising for at least 13 of them to have randomly escaped to the mother's voice but not to the conventional alarm.

Odd-Numbered Solutions to Additional Topics Exercises

Topic 33 Solutions

Homework Exercises

Exercise 33-7: Comparing Diet Plans

Let μ_1 represent the population mean adherence score for the population of dieters on the Atkins diet.

Similarly for μ_2, μ_3 and μ_4.

You want to test H_0: $\mu_1 = \mu_2 = \mu_3 = \mu_4$ (no difference in average adherence on the 4 diets).

H_a: at least one diet has a different mean adherence level after 12 months

Technical conditions: Assessing the technical conditions, you do have random assignment to the four diets. The sample distributions show a definite positive skew:

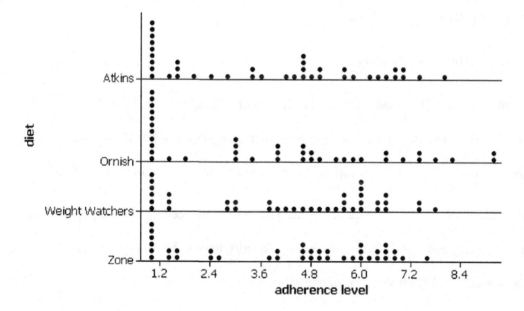

The sample sizes are reasonably large and the shapes and standard deviations (2.285, 2.590, 2.141, 2.172) are similar. You can consider the equal standard deviation condition to be met. Using technology, you have the following output:

```
Analysis of Variance for adherence level
```

```
Source    DF      SS      MS      F       P
```

diet	3	7.152	2.384	0.45	0.718
Error	156	828.090	5.308		
Total	159	835.242			

Because the *p*-value is not small (.718 > .05), you fail to reject the null hypothesis. You do not have convincing evidence that any of the diets lead to smaller or larger adherence scores, on average. Considering the technical conditions, it might make sense to re-analyze these data using only the dieters who fully completed the program.

Exercise 33-9: Sleeping Times

a. H_0: $\mu_1 = \mu_2 = \mu_3$, where μ_i is the mean sleep time for the "population" of section 1 students, etc. You are assuming the students in this study can be treated as a representative sample of a larger population of 7AM statistics students and so on.

   H_a: at least one μ_i differs from the others

b. Ask for student conjecture. For example, there is a fair bit of overlap between the three distributions, but also a distinct shift to the right with each later section. While the sample sizes are not huge, because of this consistent pattern, the *p*-value will probably be moderately small.

c. You would conclude there is a difference in the mean sleep time between these three populations, but you can't draw any cause and effect conclusions. You also have to be very cautious in thinking about what populations you can generalize to.

d. If both the lower and upper endpoints of the confidence interval have the same sign, that indicates a statistically significant difference. The output provided indicates that section 1 and section 2 are not significantly different, nor are section 2 and section 3. There is a significant difference between section 1 and section 3.

Exercise 33-11: Backpack Weights

a. H_0: $\mu_{male} = \mu_{female}$ vs. H_a: $\mu_{male} \neq \mu_{female}$

Using technology, $t = 1.05$, p-value = .298 (df = 95)

```
sex      N    Mean    StDev   SE Mean

Female  55   0.0806  0.0374   0.0050

Male    45   0.0729  0.0356   0.0053
```

```
Difference = mu (Female) - mu (Male)

Estimate for difference:  0.00767

95% CI for difference:  (-0.00687, 0.02221)

T-Test of difference = 0 (vs not =): T-Value = 1.05  P-Value = 0.298  DF = 95
```

You would fail to reject the null hypothesis. You do not have convincing evidence that the population mean ratio differs between males and females at Cal Poly.

b. Using technology, you have the following output:

```
Analysis of Variance for ratio


Source  DF        SS         MS      F      P
sex      1   0.001456   0.001456   1.09   0.300
Error   98   0.131431   0.001341
Total   99   0.132887
```

c. The test statistic values are similar (1.05 vs. 1.09) and the p-values are similar (.298 vs. .300). In fact, you should compare $t^2 = 1.05^2 = 1.10$ to $1.09 = F$. (The correspondence would be exact if you had used a "pooled" t test in part a.)

d. Both procedures require you to consider these as independent random samples from the population of interest. The students did attempt to select their participants at random and the behavior of the males should not be related to that of the females.

Dotplots reveal that both distributions are reasonably normal with moderate sample sizes, so you will consider that condition met for both procedures.

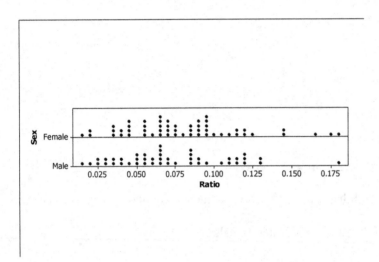

The standard deviations are also quite similar (.0374, .0356), so you will consider that additional condition met for the ANOVA. In fact, if you made this same assumption with your two-sample t-test, your p-values would match exactly.

Exercise 33-13: Memorizing Words

a. The observational units are the people.

b. The explanatory variable is *type of words* (categorical with 4 categories) and the response variable is *how many words were remembered* (quantitative).

c. This is an experiment because the list given to the subjects was randomly assigned by the students.

d. The graphical and numerical summaries follow:

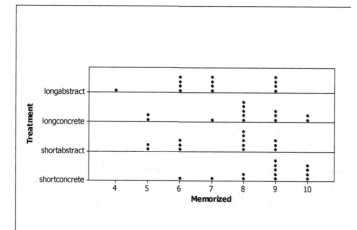

```
Variable    treatment        N    Mean    StDev   Median   IQR
memorized   longabstract     13   7.077   1.553   7.000    3.000
            longconcrete     13   8.000   1.581   8.000    1.500
            shortabstract    13   7.308   1.494   8.000    2.500
            shortconcrete    13   8.769   1.235   9.000    2.000
```

There is a lot of overlap in the four distributions of number of words memorized. There appears

to be a tendency for fewer words to be remembered with the abstract lists, and perhaps a slight

tendency for more words to be remembered with the short lists. The spread and shape of the data

also appear similar (shape is a little hard to discern, perhaps skewed to the left).

e.   Let μ_{LA} represent the treatment mean number of words remembered with the list of long, abstract

words. Similarly for μ_{LC}, μ_{SA}, and μ_{SC}.

H_0: $\mu_{LA} = \mu_{SA} = \mu_{LC} = \mu_{SC}$ (the underlying treatment means are the same for the four treatments)

H_a: at least one treatment results in a different average number of words remembered

The standard deviations are reasonably similar (1.581 /1.235 < 2) and this was a randomized

experiment.

Technical conditions: It is fairly questionable whether the treatment distributions follow a normal

distribution. Because the sample sizes are moderate, you should proceed with caution.

Here is the Minitab output:

One-way ANOVA: memorized versus treatment

```
Source      DF      SS      MS      F       P
treatment    3    22.67    7.56   3.49   0.023
Error       48   104.00    2.17
Total       51   126.67

S = 1.472    R-Sq = 17.90%    R-Sq(adj) = 12.77%
```

An F statistic of 3.49 is moderately surprising if the null hypothesis is true based on randomization alone. With a p-value of .023, you would reject the null hypothesis at the 5% level but not the 1% level. There is some evidence that the treatment mean number of words memorized differ depending on the type of list. But this p-value may be inaccurate because the treatment distributions do not appear to be normal.

f.   You can draw a cause and effect conclusion because this is an experiment, which applies to students on this campus (you have to be cautious about generalizing to this larger population because the sample was chosen by convenience).

Exercise 33-15: Cemetery Ages

a.   The boxplots reveal the ages at St. Patrick's tend to be younger but also quite variable. The distribution of ages is most consistent at the Cambria cemetery, but there are also four low outliers.

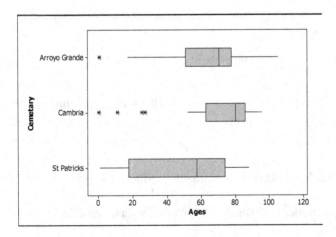

b. Let μ_{AG} represent the mean age of death for all graves in the Arroyo Grande cemetery. Similarly for μ_C and μ_{SP}. You want to test H_0: $\mu_{AG} = \mu_C = \mu_{SP}$ vs. H_a: The mean age differs for at least one of the cemeteries.

Technical conditions: The data were collected as independent random samples from each cemetery.

The largest SD ≈ 30.15 and the smallest SD ≈ 24.28, so the ratio is $30.15/24.28 < 2$.

You do have some concerns about normality, especially with the St. Patrick's cemetery, but the sample sizes are moderately sized so you will proceed with caution. The dotplots follow:

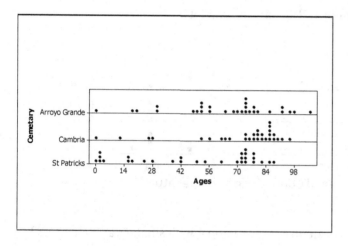

Here is the output:

```
Source   DF     SS     MS      F       P

Factor   2      7212   3606    5.05    0.008

Error    87     62133  714

Total    89     69345
```

With the small p-value ($.008 < .05$), reject the null hypothesis and conclude that at least one of the cemeteries does have a different population mean age.

c. This is a plausible explanation because life expectancies in general have increased over time.

d.   Let μ represent the mean age at death for everyone buried in these two cemeteries. You want to

test H$_0$: $\mu = 77$ vs. H$_a$: $\mu > 77$.

However, you find $\bar{x}$ for the 60 graves combined to be 66.83. Because this is lower than 77, you

will fail to reject the null hypothesis.

Exercise 33-17: Weather Predictions

a.   You can compare the high temperatures predicted by each source using ANOVA, although it is

important to note that these observations are not independent because they are grouped by day. A

more complete analysis would include the day variable in the analysis as well (similar to a

matched pairs analysis with just two groups). Here are the dotplots:

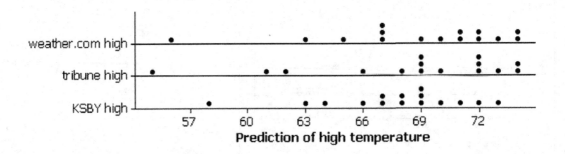

You find that the temperature predictions are reasonably normal with similar variability (4.82,

5.40, and 3.80 degrees). The means are also similar (68.73, 68.40, and 67.6 degrees) with KSBY

tending to have slightly smaller predictions.

This last observation is supported by the ANOVA output, which results in a large *p*-value.

```
Source   DF    SS    MS     F      P

Factor    2   10.2   5.1   0.23   0.796

Error    42  934.1  22.2

Total    44  944.3
```

The difference in the mean predicted high temperature is not statistically significant.

b.	Now you want just a one-sample t-test comparing μ, the prediction errors of the high

temperatures by the local television station, to 0.

H_0: $\mu = 0$ (on average, prediction error is zero)

H_a: $\mu \neq 0$ (tendency to over-estimate or to under-estimate)

The distribution is fairly symmetric with mean $\bar{x} = -1.33$ degrees and standard deviation $s = 3.02$

degrees. Here is the dotplot:

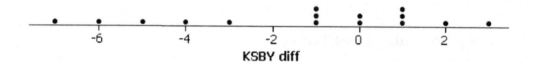

If you consider the population of prediction errors to follow a normal distribution, then you can

apply the one-sample t-test. Using technology, you find $t = -1.71$ and p-value $= .109$.

At the 10% level of significance, you do not have convincing evidence that $\mu \neq 0$. You are 95%

confident that μ falls between -3.00 degrees and $.337$ degrees, which (just) includes zero.

However, you should note that you have some evidence of a tendency to under-predict. It would

be worthwhile to collect a new sample for this source and test for a tendency to under-estimate

the high temperature.

Exercise 33-19: Texting Use

a.	Let μ_{QWERTY} represent the mean number of texts reported per day by all Cal Poly students with

cell phones with a QWERTY style keyboard. Let $\mu_{standard}$ represent the mean number of texts

reported per day by all Cal Poly students with cell phones with a standard style keyboard.

H_0: $\mu_{QWERTY} = \mu_{standard}$ and H_a: $\mu_{QWERT} \neq \mu_{standard}$

(*Note:* The ANOVA p-value assumes a two-sided alternative hypothesis.)

b. With a p-value $= .021 < .05$, the students should reject the null hypothesis. These samples provide

convincing evidence that the average number of texts reported depends on the type of keyboard

(for all Cal Poly cell phone users).

c. $$t = \frac{33.21 - 21.60}{\sqrt{15.39^2/14 + 15.97^2/42}} = 2.42$$

With df $= 13$, p-value $\approx 2 \times .015 = .030$.

With df $= 23$, p-value $\approx .024$ (using Minitab).

You expect the t-test statistic squared to approximately equal the F test statistic $(2.42^2) = 5.86 \approx$

5.65, and the p-values to be similar (.024 vs. .021). (The correspondence is not exact due to

rounding and the ANOVA procedure pooling the standard deviations.)

d. Because you have more than 10 successes (with a QWERTY keyboard) and 10 failures (with a

standard keyboard), you can apply the one-sample x confidence interval (see page 314).

$\hat{p} = 14/56 = .25$

$.25 \pm 1.96\sqrt{.25(.75)/56} = .25 \pm .11$

You are 95% confident that between 14% and 36% of all Cal Poly cell phone users have a

QWERTY keyboard.